HYDRAULIC POWER SYSTEMS

HYDRAULIC POWER SYSTEMS

By

Taruna Bhatiya

2016

SBS Publishers & Distributors Pvt. Ltd.
New Delhi

ISBN 13 : 9789380090832

First Published in 2016

Published by:

SBS PUBLISHERS & DISTRIBUTORS PVT. LTD.

2/9, Ground Floor, Ansari Road, Darya Ganj,

New Delhi - 110002,

INDIA

Tel: 0091.11.23289119 / 41563911

Email: mail@sbspublishers.com

www.sbspublishers.com

Preface

Hydraulic power, also called fluid power, power transmitted by the controlled circulation of pressurized fluid, usually a water-soluble oil or water–glycol mixture, to a motor that converts it into a mechanical output capable of doing work on a load. Hydraulic power systems have greater flexibility than mechanical and electrical systems and can produce more power than such systems of equal size. They also provide rapid and accurate responses to controls. As a result, hydraulic power systems are extensively used in modern aircraft, automobiles, heavy industrial machinery, and many kinds of machine tools. Motors in a hydraulic power system are commonly classified into two basic types: linear motors and rotational motors. A linear motor, also called a hydraulic cylinder, consists of a piston and a cylindrical outer casing. The piston constitutes the mechanical interface across which kinetic energy from the fluid is transferred to the motor mechanism. A piston rod serves to couple the mechanical force generated inside the cylinder to the external load. Hydraulic linear motors are useful for applications that require a high-force, straight-line motion and so are utilized as brake cylinders in automobiles, control actuators on aircraft, and in devices that inject molten metal into die-casting machines. A rotational motor, sometimes called a rotary hydraulic motor, produces a rotary motion. In such a motor the pressurized fluid supplied by a hydraulic pump acts on the surfaces of the motor's gear teeth, vanes, or pistons and creates a force that produces a torque on the output shaft. Rotational motors are most often used in digging equipment, printing presses, and spindle drives on machine tools. Hydraulic systems were utilized for brake systems on early aircraft.

Editor

Contents

Chapter 1

FRACTURES AND FRACTURING: HYDRAULIC FRACTURING IN JOINTED ROCK

Charles Fairhurst[1, 2]

[1] Senior Consultant, Itasca Consulting Group, Inc, Minneapolis Minnesota, USA

[2] Professor Emeritus, University of Minnesota, Minneapolis, Minnesota, USA

Rock in situ is arguably the most complex material encountered in any engineering discipline. Deformed and fractured over many millions of years and different tectonic stress regimes, it contains fractures on a wide variety of length scales from microscopic to tectonic plate boundaries.

Hydraulic fractures, sometimes on the scale of hundreds of meters, may encounter such discontinuities on several scales. Developed initially as a technology to enhance recovery from petroleum reservoirs, hydraulic fracturing is now applied in a variety of subsurface engineering applications. Often carried out at depths of kilometers, the fracturing process cannot be observed directly.

Early analyses of the hydraulic fracturing process assumed that a single fracture developed symmetrically from the packed off-pressurized interval of a borehole in a stressed elastic continuum. It is now recognized that this is often not the case. Pre-existing fractures can and do have a significant influence on fracture development, and on the associated distributions of increased fluid pressure and stresses in the rock. Given the usual lack of information and/or uncertainties concerning important variables such as the disposition and mechanical properties of pre-existing fracture systems and properties, rock mass permeabilities, in-situ stress state at the depths of interest, fundamental questions as to how a propagating fracture is affected by encounters with pre-existing faults, etc., it is clear that design of hydraulic fracturing treatments is not an exact science.

Fractures in fabricated materials tend to occur on a length of scale that is small; of the order of the 'grain size' of the material. Increase in the size of the structure does not introduce new fracture sets.

Numerical modeling of fracture systems has made significant advances and is being applied to attempt to assess the extent of these uncertainties and how they may affect the outcome of practical fracturing programs. Geophysical observations including both micro-seismic activity and P- and S-wave velocity changes during and after stimulation are valuable tools to assist in verifying model predictions and development of a better overall understanding of the process of hydraulic fracturing on the field scale. Fundamental studies supported by laboratory investigations can also contribute significantly to improved understanding.

Given the widening application of hydraulic fracturing to situations where there is little prior experience (e.g., Enhanced Geothermal Systems (EGS), gas extraction from 'tight shales' by fracturing in essentially horizontal wellbores, etc.) development of a greater understanding of the mechanics of hydraulic fracturing in naturally fractured rock masses should be an industry-wide imperative. HF 2013 International Conference for Effective and Sustainable Hydraulic Fracturing is very timely!

This lecture will describe examples of some current attempts to address these uncertainties and gaps in understanding. And, it is hoped, it will stimulate discussion of how to achieve more effective practical design of hydraulic fracturing treatments.

INTRODUCTION

The term 'rock' covers a wide variety of materials and widely different rheological properties often proximate to each other in the subsurface. Tectonic and gravitational forces, sustained over millions of years, have deformed and fractured the rock on many scales. These forces are transmitted in part through the solid skeleton of the rock, and in part through the fluids under pressure in the pore spaces. Long-term circulation through rock at high temperatures at depth involves dissolution and precipitation along the fluid pathways, producing changes in the chemical composition of the fluids and modifying the overall fluid circulation.

Rock in situ is 'pre-loaded' and in a state of changing equilibrium. Any engineering activity changes this equilibrium (see Appendix 1). Often the changes can be accommodated in stable fashion, but serious instabilities can develop.

The rock mass is opaque. Although geophysics is making impressive advances in defining large structures such as faults and bedding planes, most of the features that influence the rock response to engineering activities remain hidden. Mining and civil engineering activities allow three-dimensional access to the underground and direct observation of smaller features such as fracture networks, but most of the newer engineering applications involve essentially one-dimensional access by borehole. Rock engineering problems fall into the 'data -limited' category, as defined by Star field and Cundall (1988), and strategies to address them must follow a different strategy than engineering problems where detailed and precise design information is available.

Faced with such complexity and lack of structural details, traditional subsurface engineering design has been guided by empirical procedures developed and refined through long experience.

Projects are now venturing well beyond current experience, and for many, 'novel' applications now considered (e.g., Enhanced Geothermal Systems, Carbon Sequestration, see Appendix 1). There is little experience, few guiding rules and very little data to guide the engineering approach.

Such obstacles notwithstanding, subsurface processes, both long-term geological and short term responses, to engineering activities do obey the laws of Newtonian Mechanics. Classical continuum mechanics has long been used to guide some aspects of design, but considerable care is required in practical application, due to the need to simplify the representation of the real conditions in order to obtain analytical solutions.

The remarkable developments in high-speed computation and associated modeling techniques over the past one to two decades provide an important new tool, which complemented by the appropriate field instrumentation, can augment the classical continuum analyses and help overcome the lack of prior experience. Some empiricism and general practical guidelines may still be useful for the design engineer, but these can and should be mechanics-informed.

This lecture attempts to illustrate the 'mechanics-informed' approach with respect to the practical application of hydraulic fracturing and related engineering procedures to rock engineering.

HYDRAULIC FRACTURING

Hydraulic fracturing first was used successfully in the late 1940's to increase production from petroleum reservoirs (Howard and Fast, 1970). The technology has evolved since and is now a major, essential technique in oil and gas production. This and other impressive oil industry developments, such as directional drilling, have attracted interest in application of these technologies to a variety of other subsurface engineering operations. Enhanced Geothermal Energy (EGS) is a notable example. Geothermal Energy is a huge resource. Commenting on the EGS resource in the USA, Tester et al. (2005), state:

"....we have estimated the total EGS resource base to be more than 13 million exajoules (EJ)

[1] Using reasonable assumptions regarding how heat would be mined from stimulated EGS reservoirs, we also estimated the extractable portion to exceed 200,000 EJ or about 2,000 times the annual consumption of primary energy in the United States in 2005. With technology improvements, the economically extractable amount of useful energy could increase by a factor of 10 or more, thus making EGS sustainable for centuries."

[2] -"At this point, the main constraint is creating sufficient connectivity within the injection and production well system

in the stimulated region of the EGS reservoir to allow for high per-well production rates without reducing reservoir life by rapid cooling."

[3] -Field experiments to extract geothermal energy from rock at depth by hydraulic fracturing were started in 1970 by scientists of the Los Alamos National Laboratory, USA. Two boreholes were drilled into crystalline rock (one 2.8 km deep, rock temperature 195°C; the other 3.5 km rock, 235°C) at Fenton Hill, New Mexico. Hydraulic fracturing was used to develop fractures from the boreholes in order to create a fractured region through which water could be circulated to extract heat from the rock. The experiment was terminated in 1992. Commenting on what was learned from the Fenton Hill study, Duchane and Brown (2002) note:

"The idea that hydraulic pressure causes competent rock to rupture and create a disc-shaped fracture was refuted by the seismic evidence. Instead, it came to be understood that hydraulic stimulation leads to the opening of existing natural joints that have been sealed by secondary mineralization. Over the years additional evidence has been generated to show that the joints oriented roughly orthogonal to the direction of the least principal stress open first, but that as the hydraulic pressure is increased, additional joints open."

This is an early indication that pre-existing fractures mass significantly affect how hydraulic fractures propagate in a rock mass.

INFLUENCE OF FRACTURES AND DISCONTINUITIES ON THE STRENGTH OF BRITTLE MATERIALS

Hydraulic fracturing can be considered as a technique to overcome the strength of a rock mass in situ, initiation and propagation of a crack through a system of pre-existing fractures, essentially planar discontinuities (e.g., bedding planes), and intact rock.

In examining the fracture propagation process, the pioneering work of Griffith (1921, 1924) is a logical point of departure. Griffith had identified planar discontinuities, or flaws, in fabricated materials as the reason why the observed technical strength of brittle materials was about three orders of magnitude lower than the theoretical inter-atomic cohesive (tensile) strength.[4] - Using an analytical solution by Inglis (1913) for the elastic stresses generated around an elliptical crack in a plate, Griffith observed that the maximum tensile stress at the tip of the crack $\sigma_t = \sigma_0 (1+ 2a/b)$, where a and b are the major and minor semi-axes of the ellipse, and as the ellipse degenerated to a sharp crack or flaw (i.e., as the ratio a/b became very high)[5] - , the stress σ_t could rise to a value high enough to reach the inter-atomic cohesive strength sufficient to cause the original crack to start to extend. But would the crack continue to extend and lead to macroscopic failure? To address this question, Griffith invoked the Theorem of Minimum Potential Energy, which may be stated as "The stable equilibrium state of a system is that for which the potential energy of the system is a minimum." For the particular application of this theorem to brittle rupture, Griffith added the statement, "The equilibrium position, if equilibrium is possible, must be one in which rupture of the solid has occurred, if the system can pass from the unbroken to the broken condition by a process involving a continuous decrease of potential energy."[6] -

Griffith's classical work has provided the foundation for the field of "Fracture Mechanics" [Knott (1973); Anderson (2005)] responsible for major continuing advances in the development of high-performance fabricated materials.

Since we will make reference later to this specific definition by Griffith, it is useful to re-state it here.

THEOREM OF MINIMUM POTENTIAL ENERGY

"The stable equilibrium state of a system is that for which the potential energy of the system is a minimum. The equilibrium position, if equilibrium is possible, must be one in which rupture of the solid has occurred, if the system can pass from the unbroken to the broken condition by a process involving a continuous decrease of potential energy."

Although much of classical Fracture Mechanics has emphasized applications to problems of Linearly Elastic Fracture Mechanics (LEFM) it is important to recognize that the theorem of minimum potential applies equally to inelastic problems.

MECHANICS OF HYDRAULIC FRACTURING

As used classically in petroleum engineering, hydraulic fracturing involves sealing off an interval of a borehole at depth in an oil or gas bearing horizon, subjecting the interval to increasing fluid pressure until a fracture is generated, injecting some form of granular prop pant into the fracture as it extends a considerable distance from the borehole into the petroleum bearing formation, and then releasing the pressure. This causes the sides of the fracture to compress onto the prop pant, creating a high-permeability pathway to allow oil and/or natural gas to flow back to the well and to the surface.

Figure 1 shows a simple two-dimensional cross-section through an idealized hydraulic fracture. The borehole injection point is at the center of the fracture, which is assumed to be a narrow ellipse that has extended in a plane normal to the direction of the maximum [7] - (least compressive) in-situ stress.

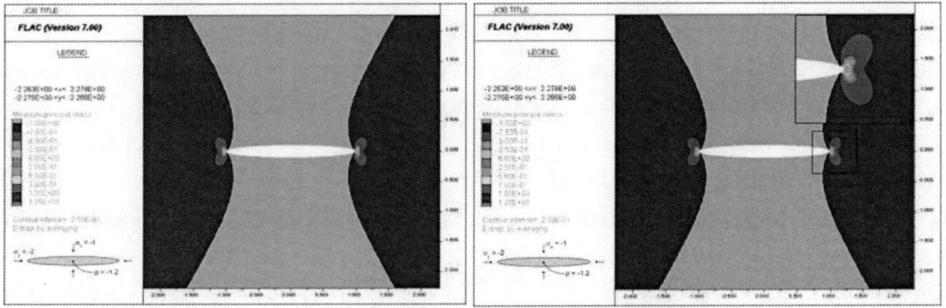

FIGURE 1. (Left) Major and (right) minor principal stresses in the vicinity of an internally pressurized elliptical crack in an impermeable rock.

In the case shown, the crack major/minor axis ratio a/b is 10:1. The internal fluid pressure p = 1.2, while the least compressive principal stress σx = 1.0. This results in a tensile stress concentration at the crack tip. The magnitude of the elastic stress concentration at the crack tip increases directly with 2a/b, (Inglis, 1913). Hence for the case of a>>b, i.e., a 'sharp' crack[8] - , the concentration is very high, and the crack will extend essentially as soon as the fluid pressure exceeds the magnitude of the least compressive principal stress (σx in Figure 3) it begins to extend, and there will be a pressure gradient from the injection point towards the crack tip as the fluid flows towards the tips. This gradient will depend on the fluid viscosity. Also, since the rock will exhibit some level of permeability, fluid will also flow (or 'leak–off') into the formation as it flows under pressure along the fracture; the rock has a finite strength, or 'toughness' so that energy will be required to extend the crack.

An analytical solution for the stresses in the elastic medium and the crack-opening displacement along the crack was first published by Inglis (1913) and served as the basis for early applications to hydraulic fracturing and fracture treatment design. The Perkins, Kern (1961) and Nordgren (1972) (PKN) and Geertsma and de Klerk (1969) (GDK) models are still used, although numerical models and combinations are now popular. Details of the PKN and GDK models can be found on the SPE website: http://petrowiki.spe.org/Fracture_propagation_models. Several differences between the stationary crack assumed by Inglis (1913) and a hydraulic fracture introduce significant difficulties in developing an accurate model of the fracturing process. Thus, the fracture is generated by application of an increasing fluid pressure until the fracture is initiated and extends away from the injection point. Flow of fluid in the fracture is governed by classical fluid flow equations of Poiseuille and Reynolds (lubrication); the pressure drop along the fracture depends on the viscosity of the fluid, and the permeability of the rock (leading to fluid 'leak-off'); the fracture aperture depends on the stiffness of the rock mass and the fluid pressure distribution along the crack; and fracture extension depends on the mechanical energy supplied to the region around the crack tip. The tip may propagate ahead of the fluid, leading to a 'lag,'a dry region between the crack tip and fluid front.

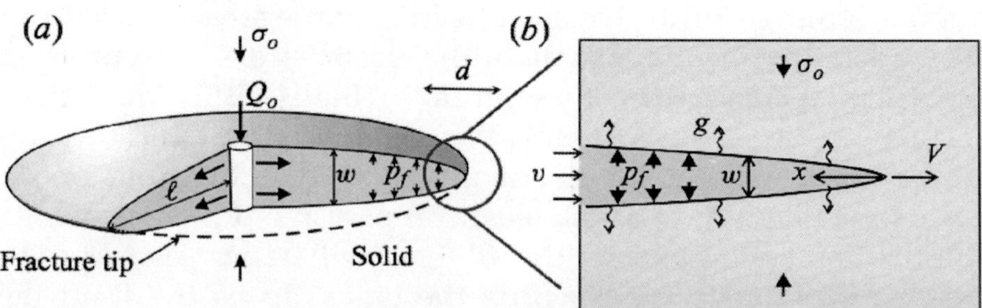

FIGURE 2. Radial Model of Axi-symmetric Flow and Deformation associated with Hydraulic Fracturing.

Figure 2 illustrates these features for the classical Radial Model in which it is assumed that the fracture propagates symmetrically away from the borehole in a plane normal to the minimum (least compressive) principal in-situ stress, σ0.

Development of efficient and robust Hydraulic Fracturing (HF) simulators is central to successful practical HF treatment of petroleum reservoirs. As noted earlier, competing physical processes are operative during the fracturing operation. This has led to a sustained effort over many years to understand and map the multi-scale nature of the tip asymptotics that arise as a result of these competing physical processes in fluid-driven fracture. These asymptotics solutions are critical to the construction of efficient and robust HF simulators. For example, in an impermeable medium, the viscous energy dissipation associated with driving fluid through the fracture competes with the energy required to break the solid material. Breaking of the bonds corresponds to the familiar asymptotic form of linear elastic fracture mechanics (LEFM), i.e., the opening in the tip region is of the form, e.g., (Rice, 1968), with denoting the distance from the tip. However, under conditions where viscous dissipation dominates, the coupling between the fluid flow and solid deformation leads to (Spence and Sharp, 1985; Lister, 1990; Desroches et al., 1994), on a scale that is considerably larger than the size of the LEFM-dominated region, but still small relative to the overall fracture size. In other words, in the viscosity-dominated regime, the zone governed by the LEFM asymptote is negligibly small compared to the crack length. Thus, in the viscosity-dominated regime, the HF simulator should embed a 2/3 power law asymptote rather than the classic 1/2 asymptote of LEFM. Garagash et al.(2011) discuss the generalized asymptotic near the tip an advancing hydraulic fracture, an extension of two particular asymptotic obtained at Schlumberger Cambridge Research Laboratory in the early 1990's (Desroches et al., 1994; Lenoach, 1995).

Three classes of numerical algorithms for HF simulators have now been built: (i) a moving grid for KGD, radial, PKN and P3D fracture simulators; (ii) a fixed grid for plane strain and axisymmetric HF with allowance for a lag between the fluid front and the crack tip, and fracture curving (a versatile code has been developed at CSIRO[9]

- Melbourne to simulate the interaction of a hydraulic fracture with other discontinuities); and (iii) fixed grid for simulating a arbitrary shape planar fracture in a homogenous elastic rock. These codes rely on the displacement discontinuity method (Crouch and Star field, 1983) for solving the elastic component of the problem, i.e., the relationship between the fracture aperture and the fluid pressure.

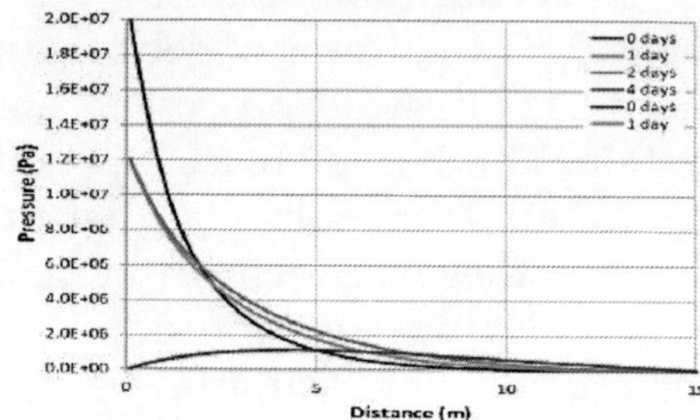

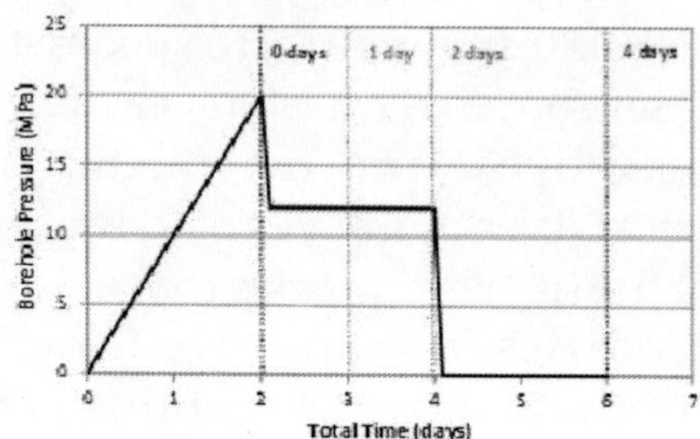

FIGURE 3. Fluid Pressure Distribution along the Central Axis (Ox) of Figure 1 for a permeable rock due to pressurization and de-pressurization of the bore-hole.

Figure 3 is presented to illustrate that the fluid pressure in a permeable rock can continue to flow away from the point of injection even after the borehole pressure is reduced to zero. The example shows the distribution of fluid pressure in the rock mass (permeability 5 mD) after (i) 2 days of pressurization up to the peak pressure of 20 MPa in the fracture; (ii) stop pumping and reduce fluid pressure quickly to 12MPa at the point of injection; (iii) hold the pressure constant for 2 days; and (iv) drop the pressure to zero.

It is seen that the pressure in the rock (red curve) has a maximum at some distance from the borehole such that fluid continues to flow into the rock for some time after the pressure in the borehole is reduced to zero. Different combinations of rock permeability, pumping rates and durations can lead to higher peak pressure values in the rock, and longer periods during which fluid can continue to flow away from the well. Such flow may contribute to slip on pre-existing fractures after the pressure in the borehole is reduced to zero.

HYDROSHEAR

Hydraulic fracturing is considered to be initiated from a packed–off interval borehole when the net state of stress around the well bore reaches the tensile strength of the rock. It is important to recognize that fluid pressurization of a well in permeable rock will result in flow of the fluid into the rock as soon as the fluid pressure stimulation process is started. This changes the effective stress state in the rock mass and can lead to slip on pre-existing fractures at fluid pressures below the pressure required to crate and extend a hydraulic fracture. This process of inducing slip on pre-existing fractures is termed 'Hydro-shear'. Flow of pressurized fluid into the rock reduces the effective normal stress (σn – p) everywhere in the rock { σn = normal stress at any point; p = fluid pressure.] If c and μ respectively represent the cohesion and coefficient of friction acting across the surfaces of a fracture in the rock, then the effective resistance of the fracture to (shear) sliding, τr, will be:

$$\tau r = c + \mu\ (\sigma n - p)$$

Thus, if the pressure p is raised progressively then τr will be reduced correspondingly until it reaches the limit at which sliding will occur. The situation is illustrated graphically in Figure 3. The rock is subjected to a three-dimensional state of stress represented by the principal stresses $\sigma 1$, $\sigma 2$, $\sigma 3$ and the fluid pressure p. The series of points 'X' indicate the effective state of stress on an array of pre-existing fractures in the rock. As illustrated in Figure 5, the effect of increasing the fluid pressure in the medium is to move the stress state on these cracks close to the limiting shear resistance, i.e., to the limiting value represented by the Mohr-Coulomb limit. As the stress state reaches this limit, the cracks will slip. In order to initiate a hydraulic fracture, the fluid pressure would need to be increased further, until the limiting Mohr circle reaches the tensile strength limit of the failure envelope. Since crack surfaces are often not smooth, shear slip will tend to result in crack dilation, and an associated increase in fluid conductivity. It is suggested that hydro-shearing could be more effective than hydraulic fracturing as a stimulation technique in certain applications, e.g., in stimulation of high-temperature geothermal reservoirs. Cladouhos et al. (2011) discuss the application of hydro-shearing as a geothermal stimulation technique. The possibility that silica prop pant may dissolve in the aggressive high-temperature fluid environment of some geothermal reservoirs whereas slip on rough fractures develops aperture increase without the need for prop pant is also presented as an argument in favor of hydro shearing.

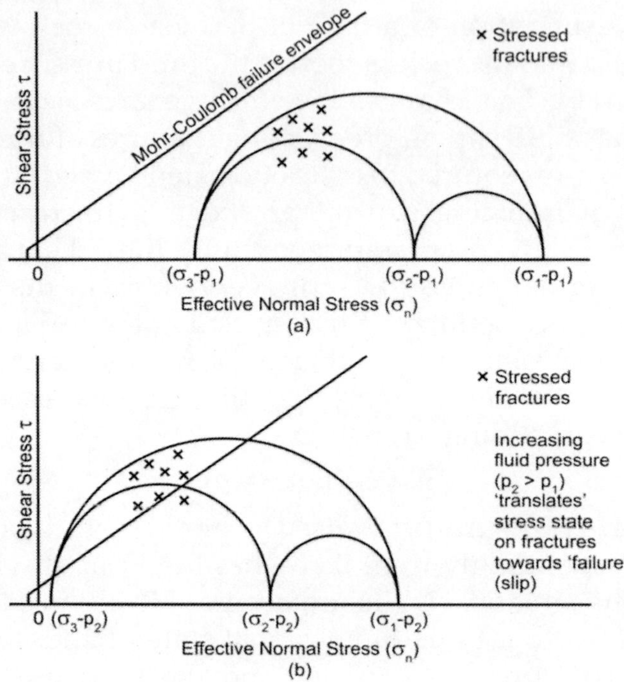

FIGURE 4. Hydro-shearing – a procedure to generate slip on pre-existing fractures by increasing the fluid pressure to a level below that required to generate a hydraulic fracture.

DEFORMATION AND FAILURE OF ROCK IN SITU

As with fabricated materials, the deformation and failure of brittle rock is also dependent strongly on fractures and discontinuities. In a rock mass, however, the fractures occur over a very wide range of scales from sub-microscopic to the size of tectonic plates. A large specimen of rock will probably include some large fractures, and as the scale of the rock mass increases, fractures from different tectonic epochs.

Study of fracture systems underground in mines and in civil engineering projects allow systems of fractures to be identified and classified statistically into discrete fracture networks (DFN's). The network will include intersecting sets of planar fractures, but individual fractures will tend to be of different lengths, and though organized in two or three spatial orientations, of variable, finite length and not collinear.

Figure 7 presents a two-dimensional illustration of the application of DFN's to the numerical modeling of a fractured rock mass. The in-situ rock mass is considered as a large specimen of intact rock that has been transected by the DFN determined from field observations and fracture mapping underground or at surface outcrops. The properties of the intact rock are built into a Bonded Particle Model of the rock (using the Particle Flow Code (PFC) code) based on results of laboratory tests of the intact rock deformability and strength. The intact rock representation is shown on the left of Figure 6. The DFN (shown on the upper right in Figure 6) then is superimposed onto the intact rock.

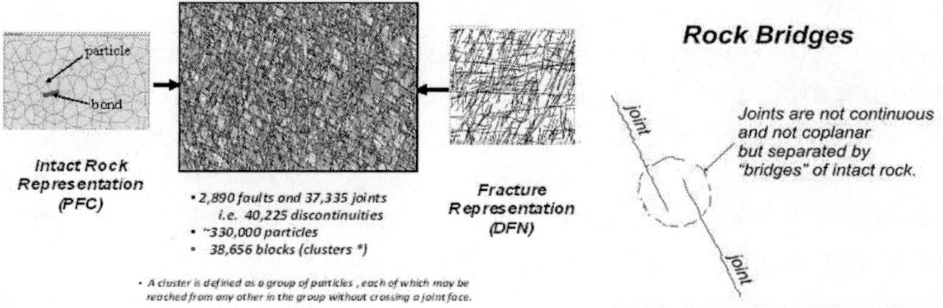

FIGURE 5. The Synthetic Rock Mass (SRM) representation of a fractured rock mass (in two dimensions). Damjanac et al. (2013) present a discussion of the 'construction' of an SRM in three dimensions. Pierce (2011) presents a comprehensive discussion of practical guidelines and factors involved in the construction of DFN's.

Cohesion and friction values are assigned to the joint planes.[10] - The 'unconfined' strength of a typical large SRM is of the order of a few percent of an intact rock specimen of the same rock (Cundall, 2008). Much of the in-situ strength is derived, of course, from the in-situ stresses imposed on the SRM in situ. One of the consequences of the finite length and lack of collinearity of joint sets in DFN's is the formation of bridges of intact rock Figure 4 within the SRM. These bridges provide regions of intact rock, and of stress concentration, in the SRM and account for a significant part of the overall strength of the rock mass. Earlier models of a rock mass, considered to consist of several sets of through-going fractures, exhibited much lower rock mass strength (Hoek and Brown, 1980).

Figure 5 presents selected extracts from a two–dimensional PFC simulation of the development of a hydraulic fracture in a jointed Synthetic Rock Mass. The SRM model was developed following the procedure outlined in Figure 5. The joint distribution was based on a DFN obtained at the Northparkes Mine in Australia.[11] - Figure 5(a) shows the location of a vertical borehole that was pressurized by fluid until a hydraulic fracture was initiated. The rock mass is assumed to be impermeable. (The path of the fracture has been traced in blue for clarity.) Displacements in the rock mass produced by the hydraulic fracture are shown as vectors on each side of the fracture. It is seen that the fracture started more or less symmetrically on each side of the borehole, but propagation of the right wing was arrested when the hydraulic fracture encountered an adversely oriented pre-existing joint (Figure 5(b)). With increasing pressure, in the borehole, the hydraulic fracture continued to extend asymmetrically towards the left (Figures 5(c) and 5(d) Figure 5(d) is simply an enlarged view of Figure. It is seen that the propagating fracture extended partially by opening existing fractures and partially by developing new fractures through intact rock. Although local deviations occur, the overall path of fracture growth is approximately perpendicular to the direction of the minimum compression stress. The existing fractures introduce an asymmetry to the rock mass. In terms of the idealized symmetric crack of Figure 2, the system in Figure 3can be considered as two cracks, one extending to the right and one to the left of the borehole with a higher 'fracture toughness' on the right compared to the left, etc.

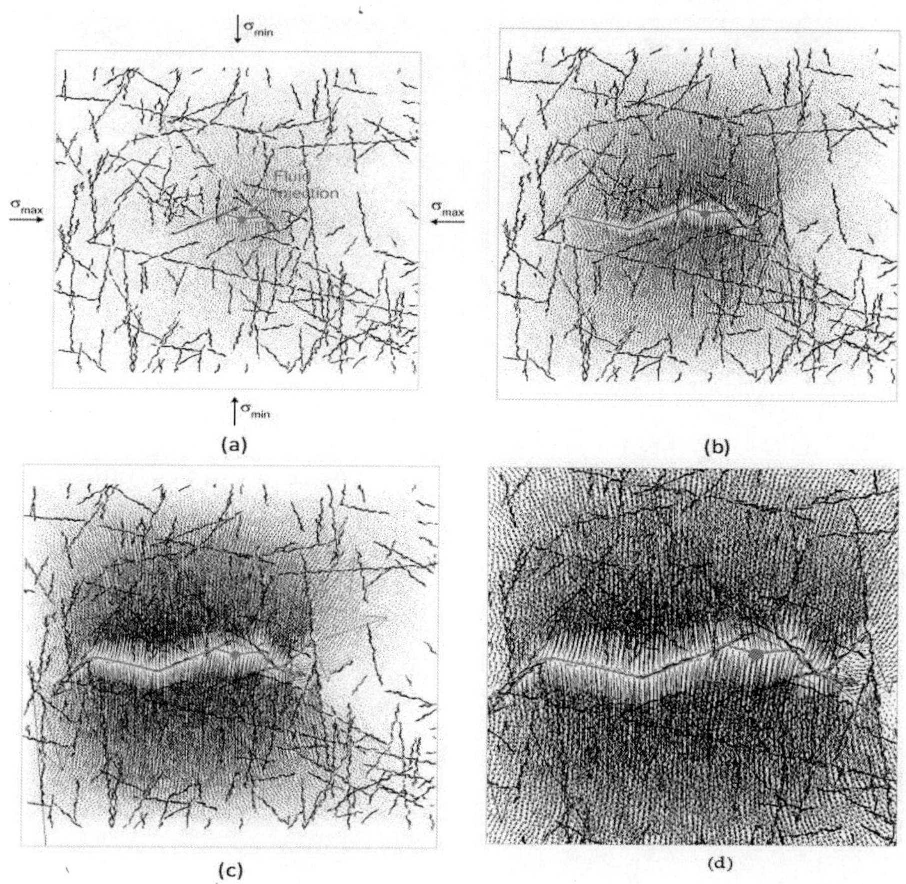

FIGURE 6. Extracts from simulation of the propagation of a hydraulic fracture in a two-dimensional impermeable SRM (Synthetic Rock Mass). (The horizontal stress σ_{max} is 29 MPa and the vertical stress σ_{min} is 12 MPa – Figure 5(a)). Note that the intact rock between the fractures has a finite strength and can break by rupture of the cemented bonded particles shown in Figure. The pressure required to propagate the fracture after breakdown was approximately 10 MPa above the minimum (i.e., least compressive) principal.

Jeffrey et al. (2009) conducted an underground test in the Northparkes Mine, Australia to observe the propagation of a hydraulic fracture in naturally fractured tock. Figure 7 shows part of the path of the fracture, as seen in a tunnel excavated into the fractured rock. The fracture path shows similar characteristics to those shown in the PFC simulation in Figure 6.

FIGURE 7. Hydraulic fracture (green plastic) crossing a shear zone on the face of a tunnel excavated through the fracture. "The arrows indicate the trace of the fracture with green plastic contained in it. There is no clear fracture between points 1 and 2 but the fracture may have crossed this zone either deeper into the rock or in the rock that has been excavated. Approximately 2 m of fracture extent is visible" (Jeffrey et al., 2009).

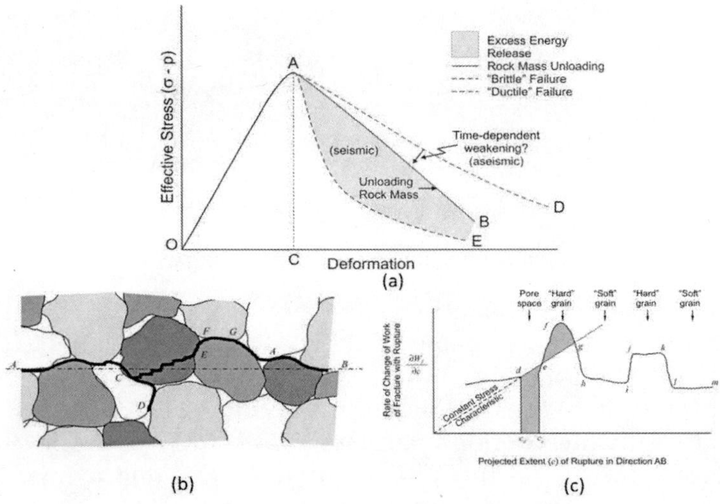

FIGURE 8. Energy changes during propagation of a fracture through heterogeneous rock.

The energy required to initiate crack propagation is represented by the area OAC in Figure 7(a). Whether or not the crack will extend depends on the energy that becomes available from the intact rock around the crack. If the energy released from the rock mass, represented by the area under the red curve AB, is greater than the energy required to extend the crack, represented by the area under curve AE, then the crack will extend; the excess energy represented by the shaded area serves to accelerate the crack and release seismic energy. If the energy required to extend the crack is represented by the area under the green curve AD, it is greater than the energy that would be released from the rock mass, and hence the crack would not extend. It is possible that the crack could exhibit some form of time-dependent weakening (e.g., due to fluid flow to the crack, viscous behavior, etc.) such that the energy required to extend the crack would be reduced. This could lead to crack extension, i.e., as the slope AD increased to overlap AB, but with no excess energy to produce seismicity. Figures 7(b) and 7(c)[12] - illustrate another feature of crack extension on the granular scale. The energy required to extend a crack through or around a grain will be variable; the fracture may encounter pore spaces where no crack energy is required. Application of a constant load to such a heterogeneous system will result in local acceleration and deceleration of the crack-producing bursts of micro seismicity. Similar effects can arise in rock fracture propagation at all scales.

It is worth noting that all of these processes of fracture propagation, albeit complex, develop in accordance with the principle of seeking the minimum potential energy of the system.

Much of the preceding discussion has focused on two-dimensional analysis or models. In reality, we are dealing with three- dimensional space (as noted in Figure 6), plus the influence of time (e.g., with respect to fluid flow, or time-dependent rock properties). Figure 8 provides an example from an actual record of hydraulic fracture propagation.

Figure 8 shows the sequence of micro seismic events observed during hydraulic fracture stimulation ('treatment' in Figure 8(a)) of a borehole. Early time events are shown as green dots; later events are in red. The micro seismic pattern indicates that fracturing started on both sides of the borehole at the injection horizon, but then moved up some 100 m to a higher horizon. As pumping continued, fracturing continued (red locations) on both horizons. It was concluded that the initial fracture in the lower horizon had intercepted a high-angle fault, allowing injection fluid to move to the higher level where it opened up and extended another fracture. Continued pumping led to fracture extension on both horizons.

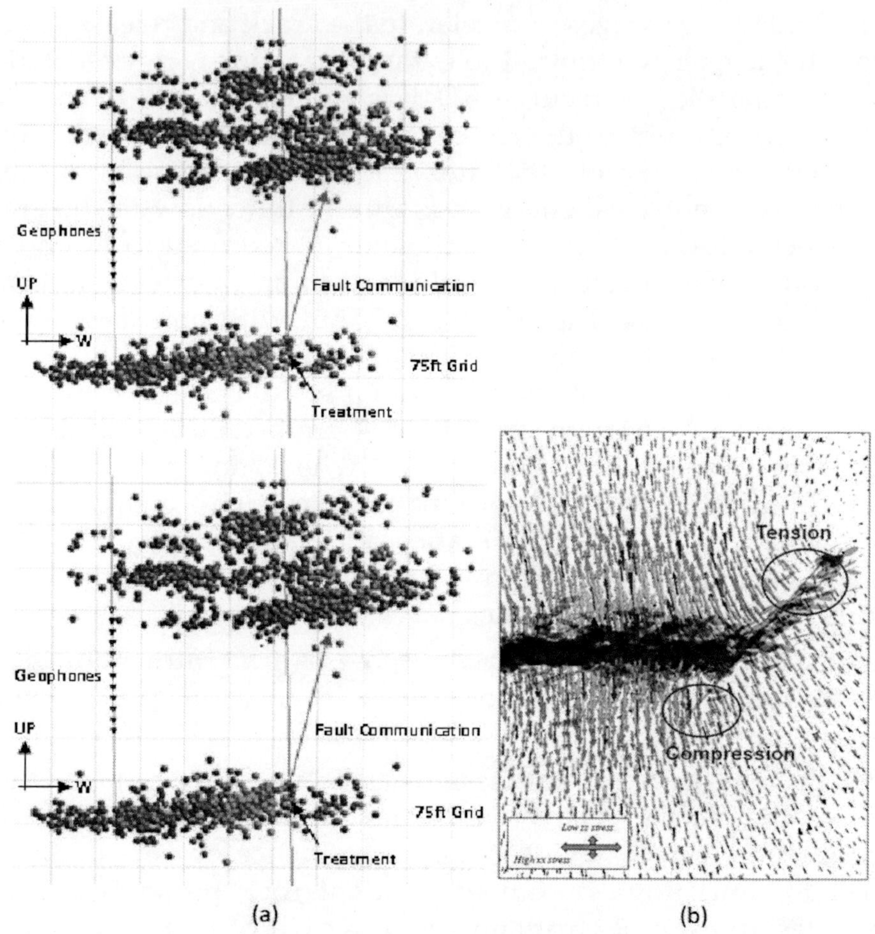

FIGURE 9. (a) Micro seismicity observed during hydraulic fracturing in a deep borehole; (b) numerical 'explanation' of the behavior observed in (a).

Numerical analysis Figure indicated that initial fracture propagation at the lower level resulted in induced tension on the fault above the horizon, but compression on the fault below the lower injection horizon. This explains why injection fluid did not penetrate along the fault below the horizon, and provides a good illustration of the benefit of combining numerical analysis with field observation in understanding fracturing processes.

MICROSEISMICITY AS AN INDICATOR OF SLIP ON FRACTURES

Micro seismicity stimulated during hydraulic fracturing and associated stimulation techniques (e.g., hydro shear) are often used to indicate slip and deformation on fractures in the rock. In some cases, it is tacitly assumed that absence of micro seismicity indicates absence of slip or deformation. In fact, there is growing evidence that micro seismicity does not present a complete picture of deformations induced by stimulation or other effects leading to stress change. Figure 9, reproduced from Cornet (2012) (with permission from the author), shows P-wave velocity changes observed by 4D (time-dependent) tomography during the stimulation of the borehole GPK2 in the year 2000. A detailed discussion of the procedure used to observe and determine the P-wave changes is presented by Calo et al. (2012).

It is seen that the region of detected micro seismicity (the cloud of black dots is small compared to the region where the P-wave velocity is reduced by as much as 20% in some regions). Some of the changes in velocity were temporary, suggesting that they may be related to temporal changes in fluid pressure; other changes appeared to be more permanent deformation that occurred a seismically.

These observations indicate that micro seismicity, although a valuable indicator of the response of a rock mass to stimulation by fluid injection, does not identify the complete region influenced by a stimulation.

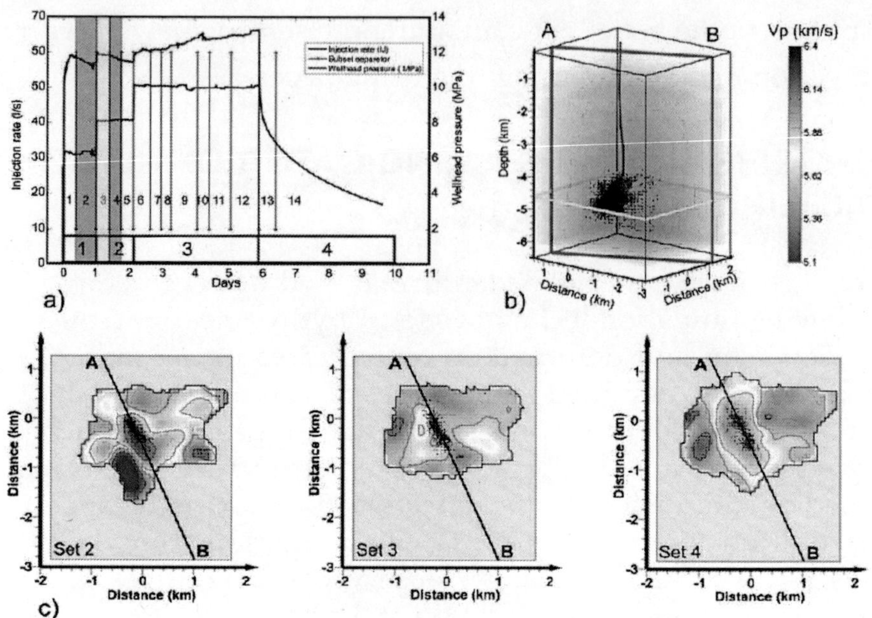

FIGURE 10. Aseismic slip induced by forced fluid flow as detected by P-wave tomography. (Soultz- sous- Fôrets, France. (a) The injection program (black curve is flow rate, blue curve is well head pressure, horizontal axis is time in days); (b) 3D view of the seismic cloud with respect to the GPK2 borehole. Vertical axis is depth and horizontal axes are distances respectively toward the north and toward the east; and (c) horizontal projections corresponding to the yellow horizontal plane. The vertical green plane is shown as line AB in the plots of part c. P-wave velocity tomography for sets 2, 3 and 4 are indicated respectively by orange, yellow and green colors in the injection program. The vertical axis corresponds to North.

IN-SITU STRESS

As already noted, hydraulic fractures tend to develop in a more or less planar fashion, extending normal to the minimum regional principal stress. Determining the direction, and perhaps the magnitude, of the regional minimum stress is an important element of hydraulic fracturing strategy, especially with the development of directional drilling, which allows borehole to be drilled in the direction considered most favorable for fracturing with respect to stress direction. (see e.g., Figure 15 and related discussion).

Determination of the in-situ stress state also can be a significant challenge.

Stress in rock is distributed throughout the mass, and is influenced by the complicated structure of the mass [13] - . Most techniques of stress determination rely on what are essentially 'point' determinations. One difficulty of determining the regional stress is illustrated by the simple, albeit somewhat artificial, example of Figure 11. This shows a two-dimensional numerical model of the stress distribution in an elastic plate containing several finite frictional fractures.

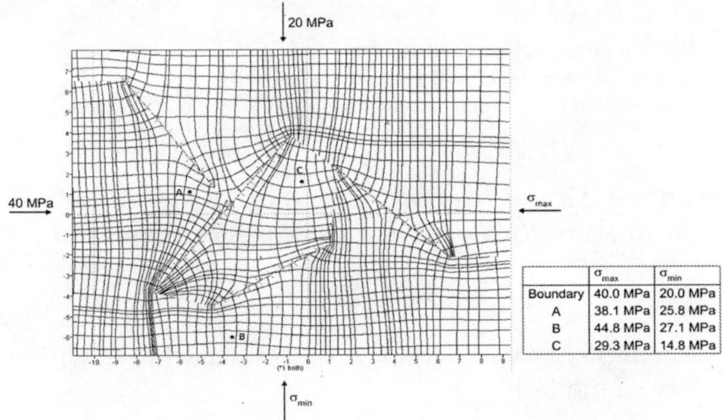

	σ_{max}	σ_{min}
Boundary	40.0 MPa	20.0 MPa
A	38.1 MPa	25.8 MPa
B	44.8 MPa	27.1 MPa
C	29.3 MPa	14.8 MPa

FIGURE 11.Influence of frictional cracks on the distribution and orientation of principal stresses, illustrative example.

The exercise serves to illustrate the difficulty of making stress determinations from local point measurements, be they in a borehole or on the surface. Stresses can change in orientation and magnitude locally due to geological in homogeneities, fractures, faults, etc., many of which may be hidden or cannot be observed from the measurement location. Although determinations made at points A and B are reasonably close to the boundary values, point C is considerably different, and the directions of principal stress, as indicated by the principal stress trajectories, can be very different from the (regional) orientations, i.e., at the model boundary.

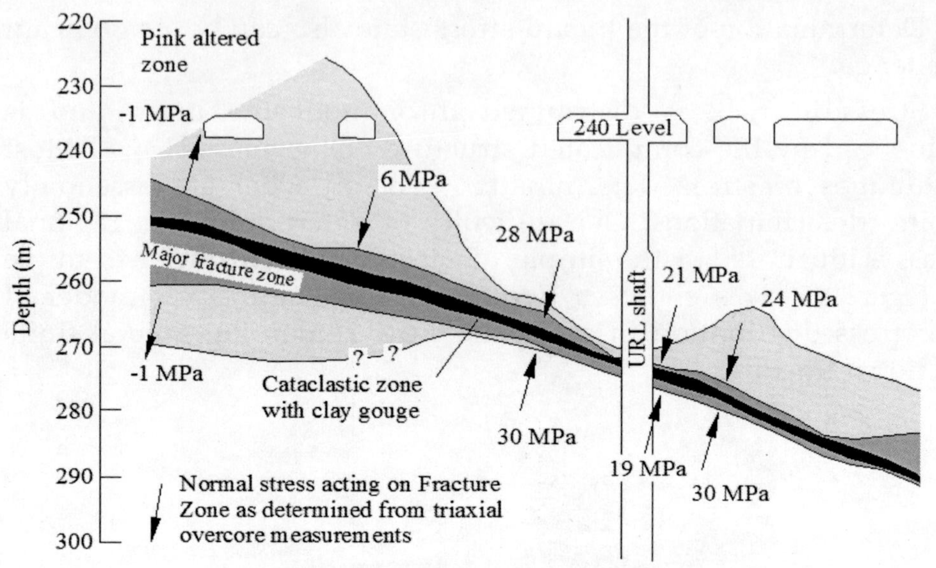

Observed Variability of Normal Stress Across a Thrust Fault Underground Research Laboratory Pinawa, Canada.

FIGURE 12. Normal stress variation across a thrust fault, Underground Research Laboratory, Canada.

Figure 12 provides an actual example of the variability of stress over relatively short distances. (The vertical and horizontal scales are equal in Figure 12). In this case, the main interest was to assess how normal stresses were affected by the thickness of gouge in the plane of the thrust fault.

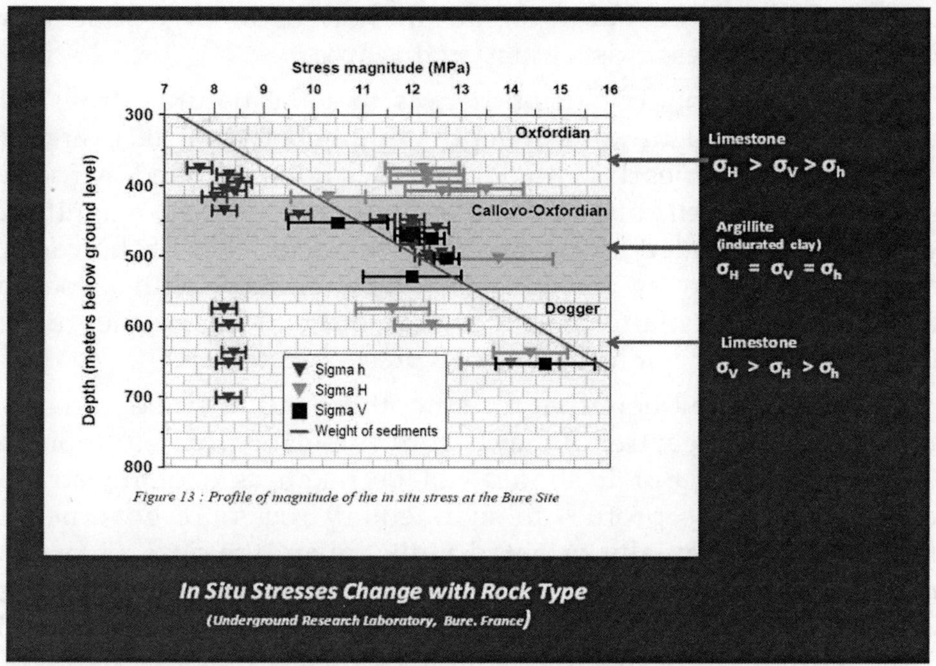

FIGURE 13.Observed stress distributions in argillite and limestone's at the Underground Research Laboratory, Bure, France.

Figure 13 illustrates another important geological influence on stress distribution, changing lithology. This example is from the French Underground Research Laboratory (URL) [14] - at Bure in NE France. Laboratory tests on specimens of the Callovo-Oxfordien Argillite indicate a long-term viscosity of this rock suggesting that any imposed deviatoric stresses would tend towards an isotropic stress state over the order of 10 million years.

Test specimens from the limestone's above and below the argillite do not appear to exhibit such viscosity. The stress distributions determined from field measurements support such differences in rheological characteristics of the rock formations.

Commenting on the in-situ stresses observations at Bure (i.e., as shown in Figure 13) Cornet (2012) notes as follows:

"Further, the complete absence of microseismicity in the Paris Basin (Grünthal and Wahlström, 2003, Fig. 4) and the absence of large scale horizontal motion as detected by GPS monitoring (Nocquet and

Calais, 2004) indicate that no significant horizontal large-scale active deformation process exists today in this area.

"The important conclusion here is that the natural stress field measured on a 100 km^2 area at depth ranging between 300 m and 700 m does not vary linearly with depth and is not controlled by friction on preexisting well- oriented faults. Rather, the stress magnitudes seem to be controlled by the creeping characteristics of the various layers rather than by their elastic characteristics, with a loading mechanism that remains to be identified but which is neither related directly to gravity nor apparently to present tectonics.

"It is concluded here that the smoothing out of stress variations with depth into linear trends may be convenient for gross extrapolation to greater depth. But it should not be taken as a demonstration that vertical stress profiles in sedimentary rocks are governed by friction along optimally oriented faults, given the absence of both micro seismicity and actively creeping fault. It should not be used for integrating together stress tensor components obtained within layers with different rheological characteristics."

Other examples could be cited, but the message is clear. Determination of in-situ stress in rock is an extremely challenging task, with results subject to considerable variability and uncertainty.

Stress orientations can be estimated from consideration of regional tectonics, faulting and interpretation of evidence from local structural geology supported in some cases by evidence based on borehole logs (e.g., tensile fractures induced along the well bore). Stress magnitudes are, in general, more difficult to determine and usually less significant, except as indicators of how stresses may be distributed across a site where the geology and engineering design are complex. In such cases, interpretation of stress distribution is best done in conjunction with a numerical model of the site, preferably one that includes the influence of important uncertainties and discussion with structural geologists familiar with the area under study.

'CRITICAL STRESS STATE' IN THE EARTH'S CRUST

It is sometimes asserted that the Earth's crust is everywhere close to a 'critical state of stress,' i.e., that a small change in the devatoric stress in the rock is likely to produce slip on one or more faults with associated seismic activity. The current global interest in development of major resources of natural gas, the central role of hydraulic fracturing in this development, and the public apprehension that hydraulic fracturing will 'trigger earthquakes' has led to strong opposition to fracturing, and even legislation to ban the use of hydraulic fracturing in some countries and some States in the USA.

As illustrated by Figure 14, the seismic hazard, (i.e., probability of a damaging earthquake) varies very considerably from place to place. Thus, an earthquake of a given magnitude is 1000 times more likely to occur in Southern California than it is in the Eastern United States. The hazard is even lower in regions such as Texas, North Dakota and in the stable Canadian Shield region of the North American tectonic plate.

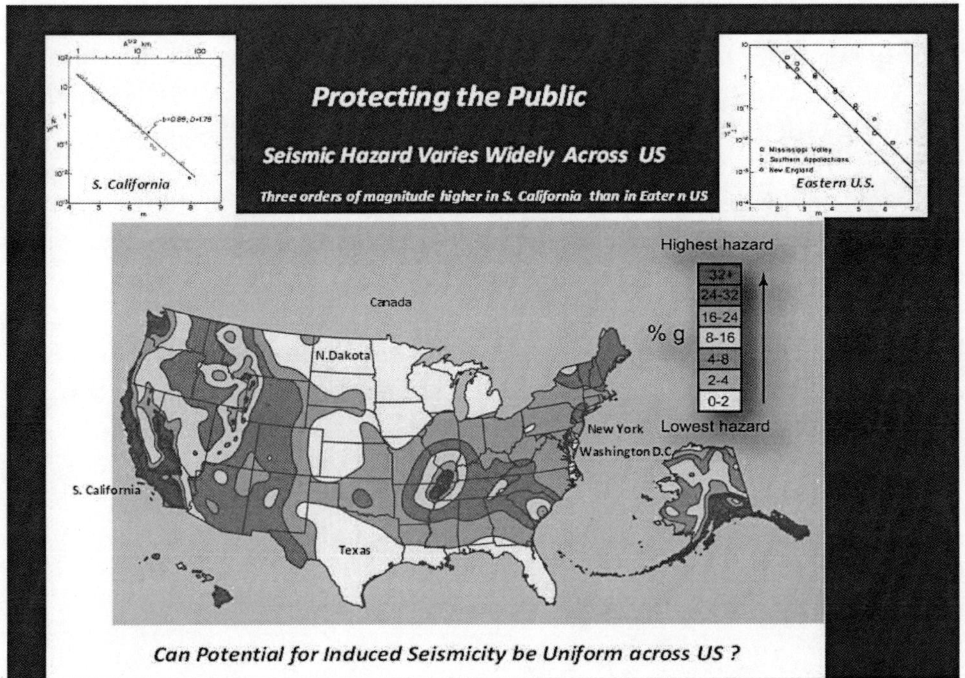

FIGURE 14. Seismic hazard map of the United States — US Geological Survey.

While many earthquakes are initiated at depths considerably greater than depths where hydraulic fracturing is applied, it seems plausible to suggest that there may be less potential for fracturing to induce seismic activity in regions that have low seismic hazard. Also, as indicated by the comments of Cornet in the previous section of this paper, there is evidence that the critical stress hypothesis warrants detailed scrutiny, at least. This could have major implications for development of the world's major natural gas and EGS (enhanced geothermal systems) resources. Two recent studies, National Research Council (2012) and Royal Society - Royal Academy of Engineering (2012), have each concluded that the risk that hydraulic fracturing as used in development of energy resources would trigger significant seismic activity is small, but it would be valuable to examine the critical stress hypothesis more rigorously than has been done to date.

HYDRAULIC FRACTURING IN TIGHT SHALES

The development of inclined and horizontal drilling (see Appendix 1 - Figure A1-2) has helped stimulate intense activity to develop natural gas production from so-called tight shale, i.e., rock in which natural gas is held tightly within the very fine pore structure of the rock. Figure 15 illustrates the procedure used to stimulate these shale's. The well is drilled horizontally in the gas-bearing formation, more or less in the direction of the minimum principal stress. Hydraulic fractures are generated (and propped) at intervals along the well to generate a network of connected flow paths that will allow the gas to flow to the well. Depth (i.e., extent) and spacing of the fractures should be optimized to produce the formations effectively. Bunger) discuss the factors in the design of an effective fracture strategy.

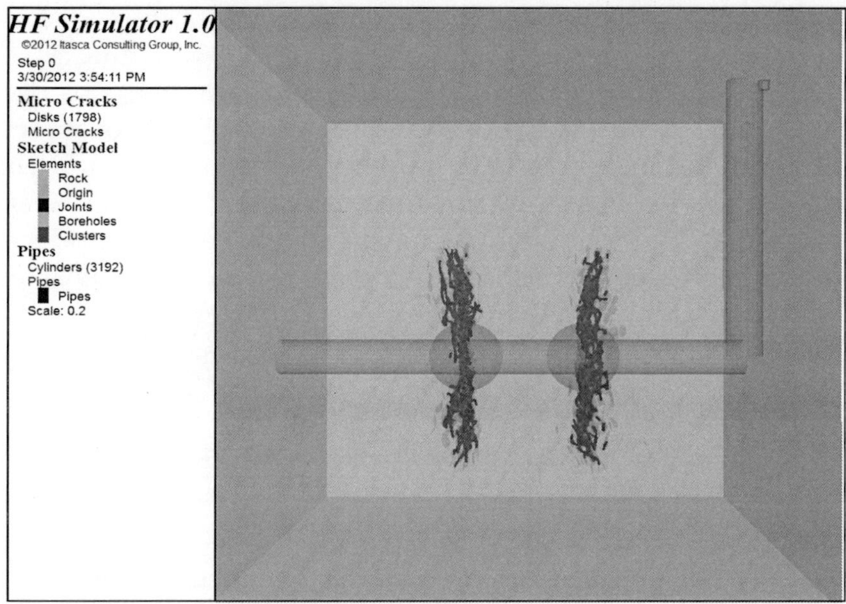

FIGURE 15. Staged hydraulic fracturing in a horizontal well. There may be many such wells along the horizontal well.

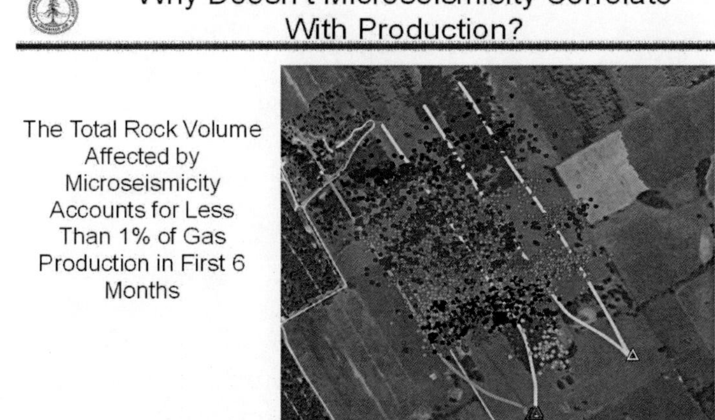

FIGURE 16. The volume of rock defined by microseismicity is a very small fraction of the volume producing gas.

Figure 16 shows a slide from a recent presentation by Prof. Mark Zoback, who kindly agreed to allow the author to include it here. Although on a somewhat smaller scale, the fact that considerable deformation and fracturing must be taking place that is not associated with detected microseismicity is similar to the phenomena discussed in connection with Figure 10. Prof. Zoback refers to such aseismic deformation as slow slip, and is conducting research to understand the underlying mechanisms, including the possible influence of the clay content of the shale. As can be seen in Figure 17 (courtesy of Prof. Zoback), the clay content can be large.

Average Shale Properties

	BARNETT	MARCELLUS	EAGLE FORD	FLOYD
Depth (ft)	3 - 9,000	2 - 9,500	4 - 13,500	6 - 13,000
TOC (%)	1 - 10	1 - 15	2 - 7	1 - 7
RO (%)	0.7 - 2.3	0.5 - 4+	0.5 - 1.7	0.7 - 2+
Porosity (%)	2 - 14	2 - 15	6 - 14	1 - 12
Qtz + Calcite (%)	40 - 50	40 - 60	50 - 80	20 - 30
Clay (%)	20 - 40	30 - 50	15 - 35	45 - 65
Areal Extent (mi²)	22,000	60,000	15,000	6,000
Resource Size (Tcf)	25 - 250	50 - 500	10 - 100	<<1

How many Floyd Shales are There?

FIGURE 17. Clay content of some typical 'tight' gas shales.

Figure 18 illustrates the very fine, micron scale, pore structure of typical tight shale. Although the mechanism(s) by which flow pathways are established in such a fine structure is not clear, the level of micro seismic energy release associated with brittle breakage of one or a few bonds will be very small and of high frequency (such that the radiated energy would be rapidly attenuated), and hence, not detectable by any geophone. Thus, absence of micro seismicity may not indicate an absence of breakage of brittle bonds. Some mechanism must be operative that generates flow pathways. Intuitively, it might be expected that the clay content of the shale might lead to ductile and viscous deformation that could tend to close the pathways.

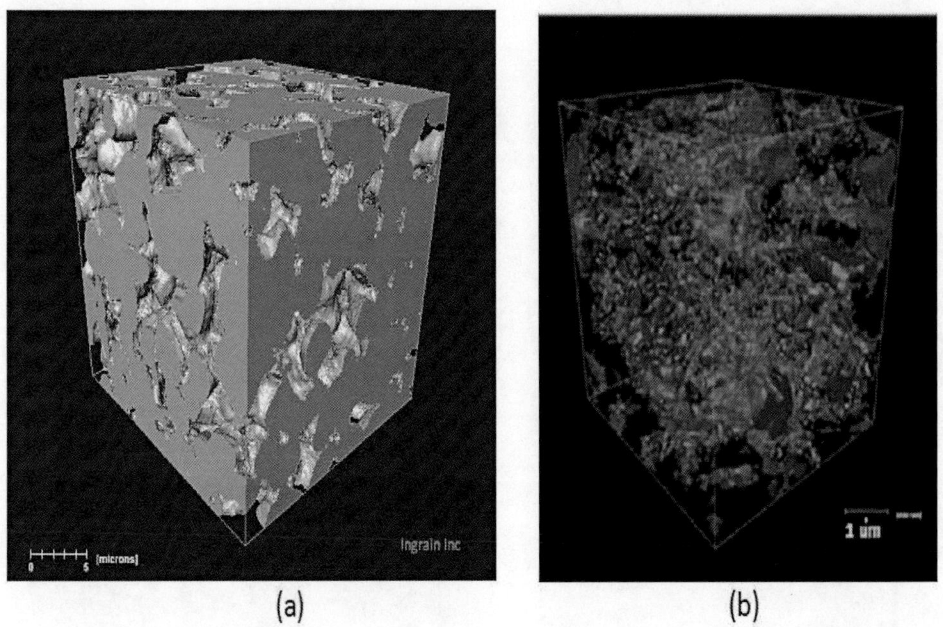

FIGURE 18. (a) Outer surface of a FIB-SEM (Focused Ion Beam- Scanning Electron Microscope) volume of Eagle Ford Shale; (b) Transparency view of the distribution of connected pores (blue), isolated pores (red) and organic matter (green). (Courtesy of Prof. Amos Nur and J. Wallis (see Wallis et al., (2012) for details of technology.)

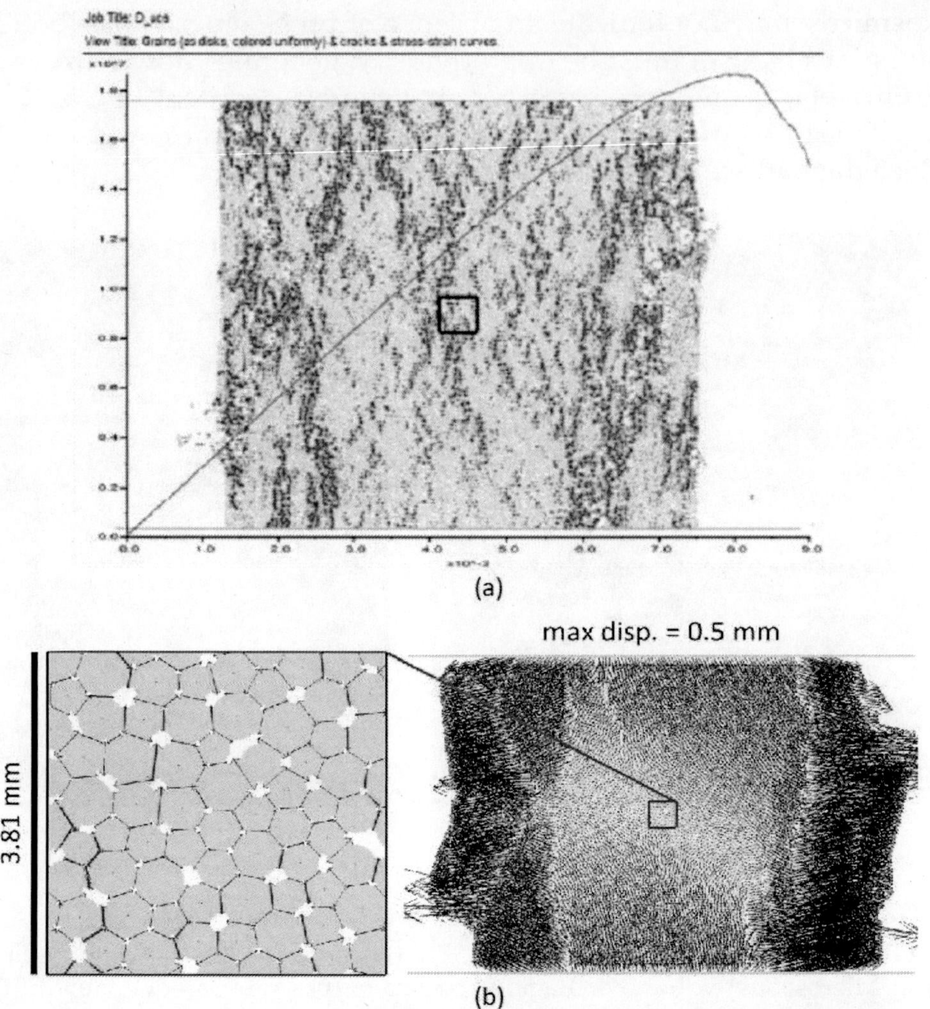

FIGURE 19. Micro-rupture of bonds within a *PFC* model of a rock loaded to failure, and beyond, in uniaxial compression. The darker red regions in (a) indicate coalescence of smaller groups of bonds that have ruptured. Eventually these larger regions develop to provide a mechanism that leads to collapse of the specimen. It is seen that bond breakage occurs throughout the specimen as the load is increased. The larger dark red regions will release larger amplitude, lower frequency waves that can be detected, whereas the smaller 'pathways' cannot be detected seismically. The load-deformation curve is shown as an 'overlay' on the specimen.

FRACTURE NETWORK ENGINEERING

This paper has emphasized the central role of fractures in rock, primarily natural fractures developed on a wide spectrum of scales over many tectonic epochs and many millions of years. These fractures and fracture systems are of special significance with respect to hydraulic fracturing and related techniques of fluid injection into rock since the fluid will tend to seek out those fractures that can be more readily opened against the local in-situ stress field as the fluid is injected. Given the complexity and lack of information on the fracture system, stress environment, etc., how can the engineering of hydraulic fracturing and related fluid injection programs advance most effectively?

Confronted with the same complexity of rock in situ, civil engineers and mining engineers have tended to adopt the 'Observational Approach' (Peck, 1969). In essence, this approach involves developing an initial engineering design for the problem, based on a first assessment/estimate of the rock (or soil) properties. Observe the actual performance and modify the initial design as needed to arrive at the desired performance. An example of the Observational Approach (as used in the New Austrian Tunneling Method) is discussed inFairhurst and Carranza-Torres (2002), see pp. 24-30.

Application of the Observational Approach to Hydraulic Fracturing and related fluid injection techniques faces some disadvantages and some advantages. We do not have 3D access to the engineering site. We do have powerful numerical modeling tools to help make a more informed initial estimate of how the system will perform; and we have sensing systems, both down hole and remote. Figure 20 illustrates a procedure that tries to apply the Observational Approach to hydraulic fracturing and related systems. The illustration describes an application to the extraction of Geothermal Energy.

Stones have begun to speak, because an ear is there to hear them. .....

Cloos, Conversations with the Earth (1954), 4

Microseismicity –predicted and observed.

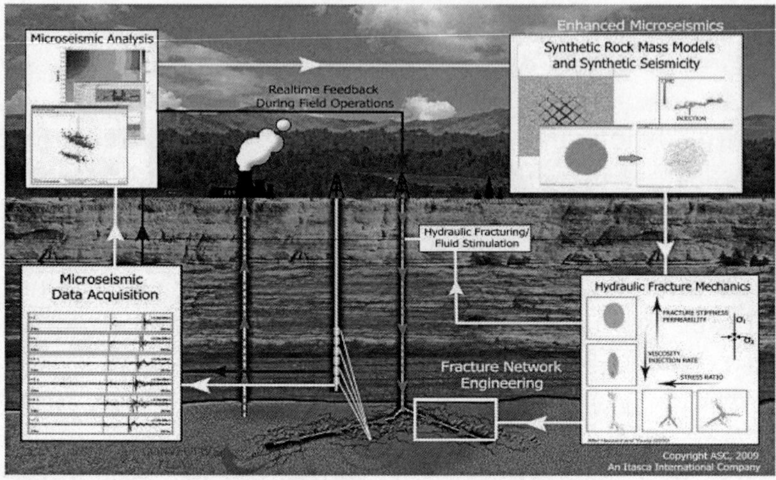

Fracture Network Engineering. *Synthetic Rock Mass and Synthetic Seismicity Models are compared with observed microseismic signals for* ***real time control of fracture network development.*** *(Enhanced Geothermal Systems.)*

FIGURE 20. Fracture network engineering system.

In this application, an initial design approach is developed based on a numerical modeling study incorporating any available data, insight, etc., on the site. This model provides an initial prediction of the performance. Instrumentation, both down hole and on-surface observes the initial response of the system and compares it with the prediction. This triggers a feedback signal to modify the design input to move the performance closer to the one desired. This iteration continues, changing progressively towards the performance desired.

Although the writer knows of no such Fracture Network Engineering system currently in operation, many of the components are available and it is time to start.

CONCLUSIONS

Expectations for higher living standards of a rising world population, and the associated demand for Earth's resources of energy, minerals and water, lead inevitably to greater focus on resources of the subsurface.

This focus includes the need to develop improved technology to develop these resources, and a better understanding of the nature of the subsurface environment as an engineering material.

Earthquakes and dynamic releases of energy are a daily reminder that on the global scale, Earth is critically stressed, and constantly trying to adjust seeking to achieve a condition of minimum potential energy for the entire system.

Ongoing for many, many millions of years, such adjustments have resulted in the heterogeneous assembly of blocks of rock bounded by essentially planar surfaces; fault, fractures and similar 'discontinuities' varying in scale from tectonic plates and continents down to micron and even nanometers.

Some of these volumes are critically stressed; others are far from a critical condition. National maps of seismic hazards provide evidence of this heterogeneity on a larger scale.

Although Earth Resource Engineering activities may be kilometers in extent, they are small-scale within the larger Earth context. Subsurface engineering in a critically stressed region can be a much different challenge than in a stable region. It is important to assess the initial conditions carefully for each case, and especially where fluid injection is a main component of a project.

The sub-surface is opaque in several ways. Details of the key features that can control the response to an engineering activity in the sub-surface are often unknown. Problems are data-limited. This is particularly the case when the engineering is based on deep borehole systems, as in hydraulic fracturing and related fluid injection technologies.

Although operating in ways that may appear complex, the response of the subsurface to stimulation does obey the laws of Newtonian mechanics, and it is clear that pre-existing natural discontinuities have a major influence on how the subsurface responds to engineered changes.

The advent of powerful computers and developments in numerical modeling provide a potentially major tool to help develop better-informed strategies of subsurface engineering. Used interactively in close conjunction with instrumentation, both down hole and surface based, it should be possible to progressively develop a mechanics-informed understanding and path forward for more effective subsurface engineering.

Much as the field of Fracture Mechanics has led, and continues to lead, to major technological improvements for fabricated materials, so can development of the field of Rock Fracture Mechanics be of transformative value to subsurface engineering, and to society in general.

Hydraulic fracturing and related injection-stimulation systems will certainly be a central element in the future of Earth Resource Engineering. The organizers of HF 2013 are to be commended for focusing attention on this critically important topic.

APPENDIX 1

Earth Resources Engineering

In 2006, the US Academy of Engineering introduced the term 'Earth Resources Engineering' to replace 'Petroleum, Mining and Geological Engineering' in recognition of the broader range of engineering activities and concerns associated with use of the subsurface. The new title, it is hoped, will also stimulate important synergies between the various disciplines involved. Mining and civil engineers, for example, have direct three-dimensional access to the subsurface not available to colleagues in other subsurface activities. This access provides a major opportunity to conduct research and gain understanding of the mechanics of subsurface processes under actual in-situ conditions, as exemplified by Jeffrey et al. (2009), see Figure A1-1.

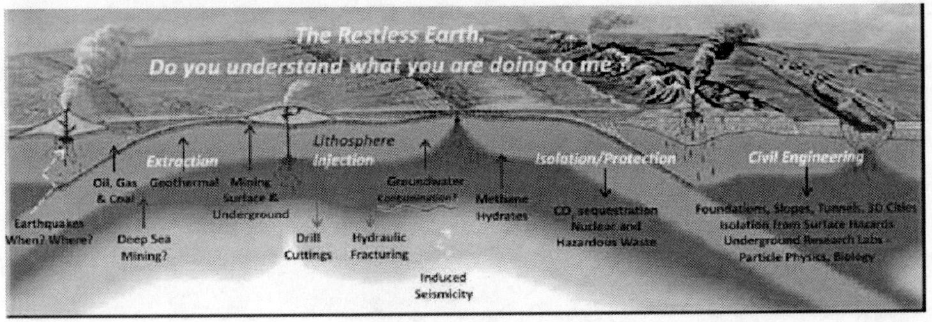

FIGURE A1-1.The restless Earth. Earth Resource Engineering activities are all confined to a very shallow part of the 40 km -700 km thick Earth's solid crust (lithosphere). Deepest borehole ~ 12 km; mine ~ 4km. Rock stress increases vertically $\sigma v \sim 27$MPa/km; laterally $\sigma h \sim (0.5\text{-}3.0).\sigma v$: Pore water pressure p = 10 MPa /km; temperature increase ~25°C /km depth.

Study of slip on active faults is a good example.

"The physics of earthquake processes has remained enigmatic due partly to a lack of direct and near-field observations that are essential for the validation of models and concepts. DAFSAM [15] - proposes to reduce significantly this limitation by conducting research in deep mines that are unique laboratories for full-scale analysis of seismogenic processes. The mines provide a 'missing link' that bridges between the failure of simple and small samples in laboratory experiments, and earthquakes along complex and large faults in the crust. There is no practical way to conduct such analyses in other environment. To unravel the complexity of earthquake processes, this project is designed as integrated multidisciplinary studies of specialists from seismology, structural geology, mining and rock engineering, geophysics, rock mechanics, geochemistry and geobiology. The scientific objectives of the project are the characterization of near-field behavior of active faults before, during and after earthquakes".[16] - See also http://www.iris.edu/hq/instrumentation_meeting/files/pdfs/IRIS_Johnston.pdf

Petroleum engineers can now reach depths in excess of 6 km and have developed advanced drilling control technologies that allow precise access to locations extending horizontally to more than 10-15 km from a single vertical hole (see Figure 2).

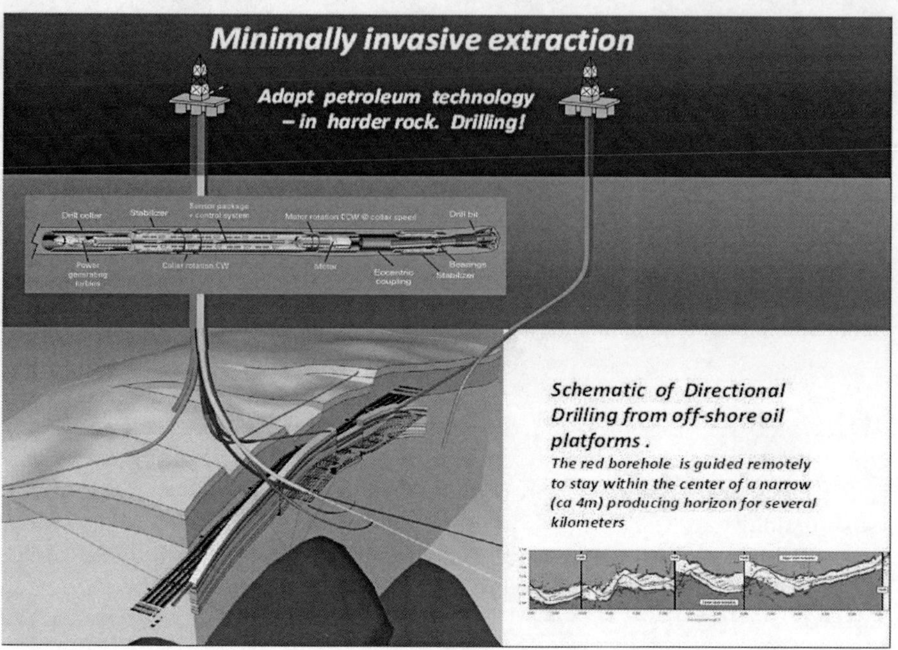

FIGURE A1-2. Schematic illustration of directional drilling for petroleum production.

These and related developments are stimulating interest in application of borehole technologies to other areas of subsurface engineering, including the development of less-invasive mining technologies, i.e., borehole extraction of minerals. Some applications, e.g., where crystalline rocks are involved, are contingent on the development of significantly lower-cost drilling technologies. The critical dependence of society on reliable and economic subsurface engineering is illustrated by the fact that currently more than 60% of the world's energy is delivered via a borehole. The Deep-water Horizon accident in the Gulf of Mexico in April 2010 provides a sober example of the consequences of error. In summary, hydraulic fracturing and related stimulation technologies are likely to see application to an increasing range of subsurface engineering challenges. HF2013, the first International Conference for Effective and Sustainable Hydraulic Fracturing, is very timely.

APPENDIX 2

Effect of Coring In Pre-Stressed Rock

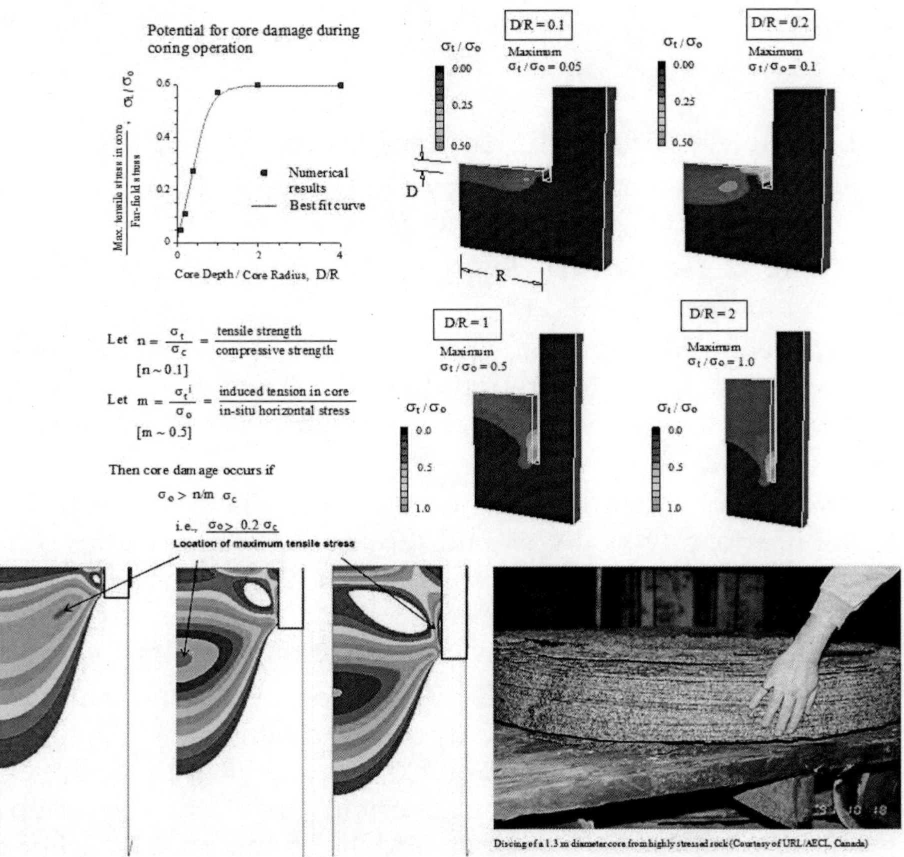

FIGURE A-2.1. Tensile stress concentrations induced in a brittle rock during coring.

The consequences of disturbing a pre-stressed rock medium are illustrated by examining the rock coring operation. Figure A2-1 shows the stress concentrations in a rock core in a brittle rock. If the in-situ stress normal to the axis of drilling is sufficiently high tensile cracks can develop in the core. Where lateral stresses are very high, then tensile 'spalling' may result, as shown in the photograph of the bottom right of Figure A2-1. Where the rock is more 'ductile' the core may undergo permanent deformation without fracturing. In both cases, the mechanical properties of these cores may differ significantly from those of the rock in situ from which the core was obtained.

Notes

[1] 1 exajoule =1018 joules = 1018 watt.seconds.

[2] Future of Geothermal Energy (2005) Synopsis and Executive Summaryp.1-4 (2).

[3] Future of Geothermal Energy (2005)Synopsis and Executive Summaryp.1-5 (5).

[4] A fractured rock mass is typically about two orders of magnitude lower in strength than the strength of a laboratory specimen taken from the rock mass [Cundall (2008); Cundall et al, (2008)].

[5] Hydraulic fractures generated in classical petroleum applications typically extend (2b) of the order of 25m ~ 50m from a wellbore. The fracture aperture (2a) at the wellbore then will be typically of the order of 0.01 m. Thus, the tensile stress concentration at the tip is very high of the order of 103.

[6] In his second paper, Griffith (1924), demonstrated that tensile stresses also developed around similar cracks loaded in compression, provided the cracks were inclined to the direction of the major principal (compressive) stress.(He also assumed that the cracks did not close under the compression.) For the optimum crack inclination, an applied compressive stress of eight times the magnitude of the tensile strength was required to develop a tensile stress on the crack boundary (close to, but not at the apex of the crack) equal to the limiting value in the tensile test. He concluded that the uniaxial compressive strength of a brittle material should be eight times greater than the tensile strength. Interestingly, he did not invoke his second (minimum potential energy) criterion. It was later determined that although a tensile crack could initiate in a compressive stress regime as predicted by Griffith (1924), the crack was

stable (i.e., did not satisfy the minimum potential energy criterion). The compressive/tensile strength ratio is greater than 8 (see Hoek and Bieniawski, 1966).

[7] Tension is assumed to be positive in Figure 3.

[8] A typical hydraulic fracture may have a length (2a) of the order of 50m and a maximum aperture (2b) of 5mm, so that the stress concentration will be of the order of 2000:1.

[9] Commonwealth Scientific and Industrial Research Organization.

[10] Typically, computer tests indicate the unconfined strength of a Synthetic Rock Mass of the order of 50-m to 100-m side length, to be a few percent of the unconfined strength of the laboratory specimen.

[11] A number of important subsurface engineering problems involve borehole access only. This often means difficulty in establishing reliable, realistic DFN's. In such cases there is no recourse, at least at the start of the project, other than to try to infer fracture networks from borehole observations, perhaps supplemented by local observations of structural geological features? The DFN for Northparkes was available and convenient to use in the example shown in Figure 5.

[12] Adapted from Fairhurst (1971).

[13] See also footnote 17 –Appendix 1.

[14] The URL at Bure was developed in order to determine the suitability of the Calllovo-Oxfordien Argillite formation for permanent storage of high–level nuclear waste.

[15] DAFSAM -Drilling Active Faults in South African Mines.

[16] http://www.icdp-online.org/front_content.php?idcat=460

ACKNOWLEDGEMENTS

Much of the material and concepts discussed in this paper is the result of work and discussions over many years with colleagues at Itasca Consulting Group, Inc. in Minneapolis and faculty in GeoEngineering at the University of Minnesota, especially in this instance, Professor Emmanuel Detournay. Particular help was received from Itasca colleagues Varun, Branko Damjanac, David Potyondy and Mark

Lorig, The influence of numerous stimulating discussions with Professor François Cornet of the Institut de Physique du Globe, Strasbourg, France are clearly evident in the paper. Professors Amos Nur and Mark Zoback, of Stanford University, USA and of Ingrain, Inc., Houston, USA assisted with valuable material, as acknowledged in the text. Dr. Rob Jeffrey and Andrew Bunger of CSIRO, Melbourne, the leaders in arranging HF2013, have provided valuable comments, assistance and understanding throughout. To all, I am very grateful. Such invaluable assistance notwithstanding, I accept full responsibility for the interpretations and views expressed in the paper.

REFERENCES

1. T. L Anderson, 2005Fracture Mechanics: Fundamentals and Applications. 3rd edition, CRC Press (0-84931-656-1
2. E. V Artyushkov, 1973Stresses in the Lithosphere Caused by Crustal Thickness
3. J Inhomogeneities, Geophy.Res. 7832November 10, 1973
4. A. P Bunger, X Zhang, and R. G Jeffrey, 2012Parameters Affecting the Interaction Among Closely Spaced Hydraulic Fractures" SPE Journal March 2012, 292306
5. M Calo, C Dorbath, F. H Cornet, and N Cuenot, 2011Large scale aseismic motion identified through 4D P-wave tomography; Geophys. J. Int. 18612951314
6. P. A Cundall, 2008An Approach to Rock Mass Modelling," in From Rock Mass to Rock Model-CD Workshop Presentations (15 September, 2008)- SHIRMS 2008 (Proc. 1st Southern Hemisphere International Rock Symposium, Perth, Western Australia, September 2008) Y. Potvin et al., Eds. Nedlands, Western Australia: Australian Centre for Geomechanics.
7. T. T Cladouhos, M Clyne, M Nichols, S Petty, W. L Osborn, and L Nofziger, 2011Newberry Volcano EGS Demonstration Stimulation Modeling" GRC Transactions, 35317322
8. F. H Cornet, 2012The relationship between seismic and aseismic motions induced by forced fluid injections." Hydrogeology Journal (2012) 20: 1463-1466
9. F. H Cornet, and T Röckel, 2012Vertical stress profiles and the significance of "stress decoupling". Tectonophysics 5812012193205

10. P. A Cundall, M. E Pierce, and D. Mas Ivars. (2008Quantifying the Size Effect of Rock Mass Strength" in SHIRMS 2008 (op.cit.) 2315

11. B Damjanac, and C Fairhurst, 2010Evidence for a Long-Term Strength Threshold in Crystalline Rock,|| Rock Mech. Rock Eng., 43, 513-531 (2010).

12. B Damjanac, C Detournay, P. A Cundall, and Varun, (2013Three-Dimensional Numerical Model of Hydraulic Fracturing in Fractured Rock Masses" Proc. HF 2013The International Conference for Effective and Sustainable Hydraulic Fracturing, Brisbane, May 20-22, 2013

13. B Damjanac, and C Fairhurst, Evidence for a Long-Term Strength Threshold in Crystalline Rock,|| Rock Mech. Rock Eng., 43, 513-531 (2010Duchane, D and D. Brown, (2002) "Hot Dry Rock (HDR) Geothermal Energy Research and Development at Fenton Hill, New Mexico" GHC (Geo-Heat Center) Bulletin, December. 2002 1319

14. C Fairhurst, and C Carranza-torres, 2002Closing the Circle- Some Comments on Design Procedures for Tunnel Supports in Rock," in Proceedings of the University of Minnesota 50th Annual Geotechnical Conference (February 2002), 2184J. F. Labuz and J. G. Bentler, Eds. Minneapolis: University of Minnesota, 2002. [available at www.itascacg.comgo to 'About'and Fairhurst Files]

15. C Fairhurst, 1971Fundamental Considerations Relating to the Strength of Rock. Colloquium on Rock Fracture, Ruhr University, Bochum, Germany, April 1971. (see http://www.itascacg.com/about/ff.php)Revised and published in Report of the Workshop on Extreme Ground Motions at Yucca Mountain, August 23-25, 2004, U.S. Geological Survey, USGS Open-File Report 20061277T. C. Hanks et al., Eds. Reston, Virginia: USGS, 2006.

16. J Geertsma, and F De Klerk, 1969A Rapid Method of Predicting Width and Extent of Hydraulic Induced Fractures. J Pet Technol 211215711581SPE-2458-PA. http://dx.doi.org/10.2118/2458-PA

17. J. F Geyer, and S Nemat-nasser, 1982Experimental Investigation of Thermally induced Interacting Cracks in Brittle Solids Int. J. Solids and Structures 184349356

18. A. A Griffith, 1921The Phenomena of Rupture and Flow in Solids Phil. Trans. R. Soc. Lond. A 1921,, 221, 163-198 doi:rsta.1921.0006

19. A. A Griffith, 1924Theory of Rupture. Proc. First Int. Cong. Applied Mech (eds Bienzo and Burgers). 5563Delft: Technische Boekhandel and Drukkerij. 1924

20. G Grünthal, and R Wahlström, 2003An Mw-Based Earthquake Catalogue for Central, Northern and Northwestern Europe using a Hierarchy of Magnitude Conversions. J. Seismol. 7, 507-531 (Available

at http://seismohazard.gfzpotsdam.de/projects/en/eq_cat/menue_e"q_cat_e.html)

21. E Hoek, and Z. T Bieniawski, 1966Fracture Propagation Mechanism in Hard Rock," in Proceedings of the First Congress of the International Society of Rock Mechanics. Lisbon, September-October, 1243249J. G. Zeitlen, Ed. Lisbon: LNEC.
22. E Hoek, and E. T Brown, 1980Underground Excavations in Rock." Inst'n of Mining and Metallurgy (London) Revised 1982, 164
23. G. C Howard, and C. R Fast, 1970Hydraulic Fracturing" SPE Monograph 2. Henry L.Doherty Series 203 pp. SPE 30402
24. C. E Inglis, 1913Stresses in a Plate Due to the Presence of Cracks and Sharp Corners," Trans. Inst. Naval Arch., London, 55(1), 219141
25. R. G Jeffrey, et al2009Measuring Hydraulic Fracture Growth in Naturally Fractured Rock. SPE 124919; SPE Annual Technical Conference and Exhibition, New Orleans, Louisiana, USA, 47October 2009
26. J. F Knott, 1973Fundamentals of fracture mechanics, Wiley (0-47049-565-0
27. National Research Council2012Induced Seismicity Potential in Energy Technologies." Washington, DC: The National Academies Press, 2012. (300p.) View online at http://www.nap.edu/catalog.php?record_id=13355
28. J. M Nocquet, and E Calais, 2004Geodetic Measurements of Crustal Deformation in the Western Mediterranean and Europe " Pure Appl. Geophy., 161; 661668
29. R. P Nordgren, 1972Propagation of a Vertical Hydraulic Fracture. SPE J. 124306314SPE-3009-PA. http://dx.doi.org/10.2118/3009-PA.
30. R. B Peck, 1969Advantages and limitations of the observational method in applied soil mechanics. Geotechnique, 192171187
31. T. K Perkins, and L. R Kern, 1961Widths of Hydraulic Fractures. J Pet Technol 139937949SPE-89-PA. http://dx.doi.org/10.2118/89-PA.
32. W. S Pettitt, J. F Hazzard, B Damjanac, Y Han, M Pierce, T Katsaga, and P. A Cundall, Microseismic Imaging and Hydrofracture Numerical Simulations," in Proceedings, 21st Canadian Rock Mechanics Symposium (Alberta, Canada, May 5-9, 2012
33. M Pierce, 2011Discrete Fracture Network Simulation" DFN training session LOP (Large Open Pit). [ppt slides available on request. Itasca Consulting Group: www.itascacg.com]
34. A Riahi, and B Damjanac, 2013Numerical Study of Interaction between Hydraulic Fractures and Discrete Fracture Networks" Proc.

HF 2013The International Conference for Effective and Sustainable Hydraulic Fracturing, Brisbane, May 20-22, 2013

35. Royal Society and Royal Academy of Engineering (2012Junep. "Shale Gas Extraction in the UK: a review of hydraulic fracturing" Issued: June 2012, DES2597] View report online at: royalsociety.org/policy/ projects/shale-gas-extraction and raeng.org.uk/shale

36. O Scotti, and F. H Cornet, 1994In-Situ Evidence for Fluid-Induced Asesismic Slip Events along Fault Zones. Int. J. Rock Mech Min.Sci. &Geomech. Abstr. 314347258Control in Mines (1965). South African Institute of Mining and Metallurgy, Johannesburg, 606p

37. A. M Starfield, and P. A Cundall, 1988Towards a Methodology for Rock Mechanics Modelling" Int. J. Rock Mech. Min. Sci.& Geomech. Abstr. 25 (3) 99106

38. J. F Tester, et al2006The Future of Geothermal Energy"-Impact of Enhanced Geothermal

39. Systems (EGS) on the United States in the 21st CenturyMIT Press.

40. J. D Wallis, J Devito, and E Diaz, 2012Digital Rock Physics- A New Approach to Shale Reservoir Evaluation" Oilfield Technology, March 2012 [http://www.ingrainrocks.com/articles/a-new-approach-to-shale-reservoir-evaluation/]

41. X Zhang, and R. G Jeffrey, 2008Re-initiation or termination of fluid-driven fractures at frictional bedding interfaces" JGR, 113BO 8416, doi:10.1029/2007JB005327,

Chapter 2

FRACTURE NETWORK CONNECTIVITY—A KEY TO HYDRAULIC FRACTURING EFFECTIVENESS AND MICROSEISMICITY GENERATION

F. Zhang[1], N. Nagel[1], B. Lee[1] and
M. Sanchez-Nagel[1]

[1] Itasca Houston, Inc., USA

ABSTRACT

In this work, the effect of fracture network connectivity on hydraulic fracturing effectiveness was investigated using a discrete element numerical model. The simulation results show that natural fracture density can significantly affect the hydraulic fracturing effectiveness, which was characterized by either the ratio of stimulated natural fracture area to hydraulic fracture area or the leakoff ratio. The sparse DFN cases showed a flat microseismic distribution zone with few events, while the dense DFN cases showed a complex microseismic map which indicated significant interaction between the hydraulic fracture and natural fractures. Further, it was found that the initial natural fracture aperture affected the hydraulic fracturing effectiveness more for the dense natural fracture case than for the sparse (less dense) case. Overall, this work shows that fracture network connectivity plays a critical role in hydraulic fracturing effectiveness, which, in-turn, affects treating pressures, the created microseismicity and corresponding stimulated volume, and well production.

INTRODUCTION

The extremely low permeability of the common shale plays means that simple, bi-planar hydraulic fractures (HF) do not create enough surface area to make economic wells and that stimulation of the natural fracture system is critical [1]. Numerous field microseismic data sets have shown that extreme fracture complexity may result from the interaction between a created hydraulic fracture and the pre-existing fracture network [2, 3]. Consequently, operators will often alter the stimulation design, by changing injection rate, viscosity, or other parameters, in order to improve the effectiveness of the stimulation in unconventional shale plays. However, these design changes offer only a limited control on improving the stimulation of natural fractures because of a lack of understanding of the fundamental characteristics, behavior, and connectivity of the natural fracture network.

The connectivity of the fracture network, for example, determines the overall hydraulic diffusivity of the formation and is a key to the resulting 'complexity' from a hydraulic fracture stimulation. A highly connected fracture network will allow more fluid leakoff into the rock mass and render pressure communication over large distances, whereas a partially or sparsely connected fracture network will favor the propagation of a new hydraulic fracture and may exhibit pressure isolation between very closely spaced hydraulic fractures. The intricacy of fluid flow in fractured formation is mainly due to the complex geometries, patterns, and heterogeneity of the fracture network. The fracture network connectivity, therefore, has been shown to be a critical factor which affects treating pressures, the created microseismicity and corresponding SRV (Stimulated Rock Volume), and production.

Numerous numerical modeling efforts have been conducted in order to understand the process of hydraulic fracture (HF) interaction with a complex natural fracture network [4, 5, 6]. However, relatively few works have focused on understanding the role of natural fracture network connectivity and its impact on the effectiveness of hydraulic fracturing of shale reservoirs and associated microseismicity generation.

In this paper, a discussion of fracture network connectivity and how it is utilized in developing a discrete fracture network (DFN) is presented, which is then incorporated into a discrete element numerical model (DEM). The propagation of a HF in the fractured rock mass was then studied using the DEM, which allowed for fully coupled, hydro-mechanical simulations, including the generation of synthetic microseismicity. Following previous work [7, 8, 9, 10] from the authors on parametric studies to analyze the influence of the mechanical and flow properties of the DFN and matrix on HF propagation, the influence of the DFN fracture network connectivity was analyzed in this work using common stimulation metrics such as SRV. The corresponding microseismic response was also calculated from the simulation results and related to the effective SRV. The results show not only the critical role that the DFN must play in resource evaluations, but also in completion design and stimulation optimization.

FRACTURE NETWORK CONNECTIVITY AND DFN REALIZATION

Fracture network connectivity is determined by many statistical characteristics, among them, fracture shape, fracture size distribution, fracture density (area of fracture per unit volume), orientation distribution, and aperture size distribution. The combinations of these statistical characteristics that describe the geometrical properties of a DFN define the macro-scale connectivity and directional flow preference of the DFN, and thus, are essential for the fluid transport characterization of an unconventional reservoir.

In this work, focus was placed on the effects of two statistical fracture characteristics, fracture density and initial aperture both individually and in combination, on hydraulic fracturing effectiveness and microseismicity generation. For the study, a DFN generator was developed that was capable of creating a fracture network that satisfied the assigned input statistical characteristics and allowed for the quantitative variation of network connectivity.

As shown in Figure 1, two DFN configurations were realized, which represent a sparse DFN and a dense DFN, respectively. The sparse DFN had 191 fractures while the dense DFN had 482 fractures, or about 2.5 times more than the sparse DFN. The fractures were created in the common disk shape with a dip angle of 90° but either a 60° or 150° dip direction in order to construct two orthogonal fracture sets. The P10 (the number of fractures along a line divided by the length of the line) for the sparse DFN and dense DFN was 0.079 m^{-1} and 0.28 m^{-1}, respectively. The P32 (the sum of the areas of the fractures contained in a named volume divided by the same volume) for the sparse DFN and dense DFN was 0.031 m^2/m^3 and 0.075 m^2/m^3, respectively.

In addition to DFN density, the effect of the initial DFN aperture was studied by considering two different values of initial aperture for the DFN fractures. In the first case, it was assumed that the initial aperture of the DFN fractures was equal to 0.1 mm. In the second case, it was assumed that the DFN fracture aperture was 1.5 times greater and equal to 0.15 mm. The goal of this study was to investigate the effects of hydraulic fracturing in shale formations with different levels of DFN connectivity. It was assumed that the DFN fractures were static, meaning that, with the exception of the main hydraulic fracture itself, the process of fracture growth and propagation was not considered within the simulations.

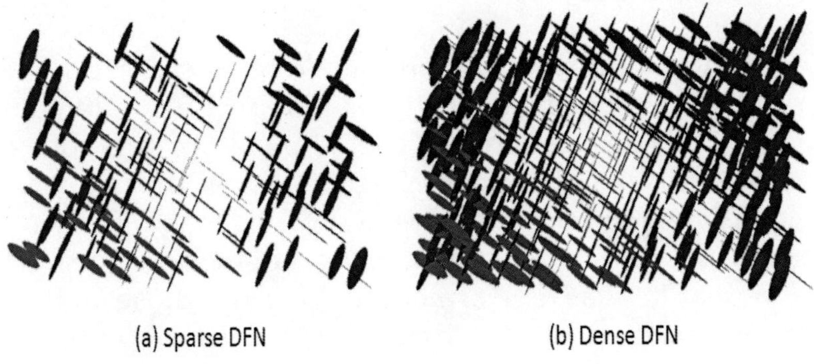

FIGURE 1. Plan view of two DFN realizations. A) Sparse DFN with 191 fractures; and B) Dense DFN with 482 fractures.

NUMERICAL SIMULATIONS

3Dec Hydraulic Fracturing Modeling Capabilities

The numerical code used in the simulations, *3DEC,* is a three dimensional distinct element method for discontinuum modeling [11]. It can simulate the response of discontinuous media, such as a jointed rock mass, subjected to either static or dynamic loading, which includes the simulation of fluid injection during a hydraulic fracture treatment. *3DEC*'s ability to handle large block displacements makes it amenable to the simulation of hydraulic fracture propagation, fracture opening (both hydraulic fracture and natural fractures), and the rotation of individual blocks leading to fracture connections or pinching. Further, *3DEC* includes the ability to model steady-state or transient-fracture fluid flow, where the flow logic includes a system of flow planes, flow pipes and flow knots. While most of the existing commercial hydraulic fracture simulators are based upon a number of analytical assumptions, or limit the simulated fractures to be smooth, bi-planar cracks, *3DEC* is capable of explicitly modeling the mechanical response of a fracture network to fluid injection as well as the interaction of a created hydraulic fracture (the HF plane needs to be pre-defined) with the existing natural fractures.

3DEC MODEL SETUP

The *3DEC* simulation domain consisted of two parts: an inner core domain with the population of the DFN and an outer boundary domain, which extended to twice of the size of the core domain in order to minimize boundary effects. To port the created natural fractures to *3DEC*, a procedure was developed to explicitly represent the generated DFN in the geomechanical model. For each DFN fracture, a search was performed to identify those *3DEC* blocks that were intersected by the given fracture. The identified blocks were then cut through into two blocks by the fracture plane. Since the newly created plane between blocks might only have part of its area belonging to the given fracture (if the natural fracture did not fully cut the blocks), different fracture material properties were assigned to the portion of the newly created plane that fell within the geological fractures than to the unfractured portion that fell outside the geological fractures.

Figure 2 shows the core simulation domain discretized by the dense DFN. Individual blocks are shown in variable color, while the plane between two blocks is a possible natural fracture. The size of the inner domain was 400×200×100 m. The directions of the three principal stresses are shown in Figure 2, which are coincident with the three axes x, y and z.

For all the simulations, the maximum horizontal stress (SHmax) was 55 MPa, the minimum horizontal stress (Shmin) was 50 MPa, and the vertical stress (Sv) was 60 MPa. The initial pore pressure was set to 45 MPa. The Young's modulus of the rock mass was set to 30.0 GPa and Poisson's ratio was set to 0.25. The injection point was located in the center of the domain along the predefined hydraulic fracture plane (the plane of y=0), which was parallel to SHmax. The injection rate was 0.05 m^3/s and the fluid viscosity was 0.0015 Pa s. It was assumed that fluid flow only occurred in either the DFN or HF planes and that there was no fluid flow into the rock matrix during the simulations.

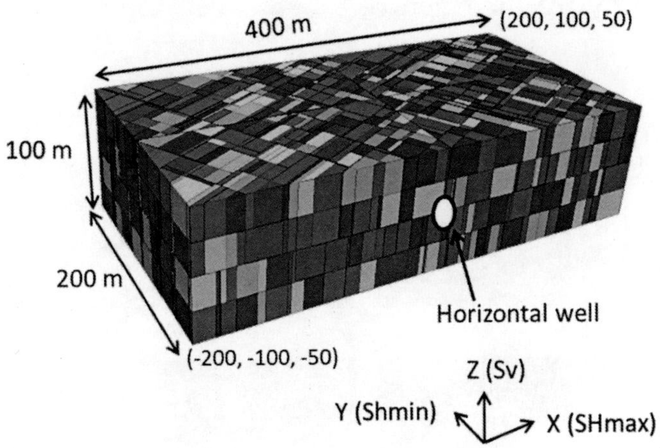

FIGURE 2. The core domain of the numerical model discretized by the dense DFN. Blocks are distinguished by colors. The plane of y=0 is the HF plane.

SIMULATION RESULTS

In this study, four simulation cases, consisting of two DFN configurations (i.e., the dense DFN and sparse DFN) and two initial DFN apertures (i.e., 0.1 mm and 0.15 mm) were analyzed. Figure 3shows a plan view of the contours of fracture pressure distribution at the XY plane of z=0 m after 900 s of injection time for the four tests.

The fracture pressure contour plot suggests two major observations. With the same initial DFN aperture, the case with the dense DFN resulted in significantly more stimulated DFN area and less stimulated HF area than the case with the sparse DFN. In addition, with the same DFN configuration, the case with 0.15 mm initial aperture resulted in more stimulated DFN area and less stimulated HF area than the case with 0.1 mm initial aperture. These results are also confirmed by the shorter hydraulic fracture lengths in the dense DFN simulations. These qualitative conclusions are as expected. Since hydraulic fracturing represents, essentially, the release (injection) of hydraulic energy into the formation, the injected fluid will follow the least resistive path. For the dense DFN, it was easier for fluid to enter the natural fractures than it was for the fluid to propagate a hydraulic fracture.

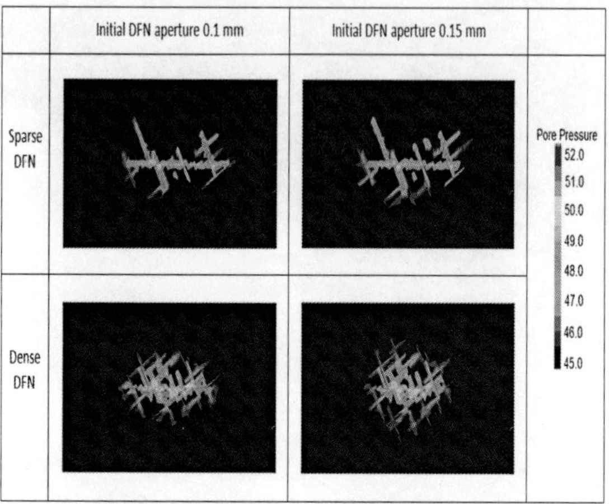

FIGURE 3. Plan view of contours of fracture pressure distribution at the XY plane of z=0 m after a 900 s injection time for the cases of sparse and dense DFN realizations with two initial DFN apertures.

The synthetic microseismic events during the injection can be approximated based on the magnitude of plastic slip. Figure 4 plots the synthetic microseismic moment magnitudes for the sparse and dense DFN with the same initial aperture (0.1 mm) after 900 s of injection time. The events that clustered based on their spatial and temporal proximity were colored by the occurring time and sized by the microseismic moment magnitude. All events were projected onto the XY plane with z=0 m and the XZ plane with y=0 m (HF plane). Figure 4 shows that the sparse DFN case had fewer microseismic events than the dense DFN case. The complexity of microseismic events for the dense DFN case was consistent with field observations and suggests an intensive interaction between the created hydraulic fracture and the natural fractures.

The two qualitative observations from Figure 3 were further proved quantitatively by tracking the evolution of stimulated DFN area and HF area. As there is no precise criteria for defining the stimulated area, a criteria based on fracture pressure change was employed in this work. The area of the DFN or HF planes having a fracture pressure increase of 0.5 MPa or 1 MPa above the initial fracture pressure was considered as the stimulated area.

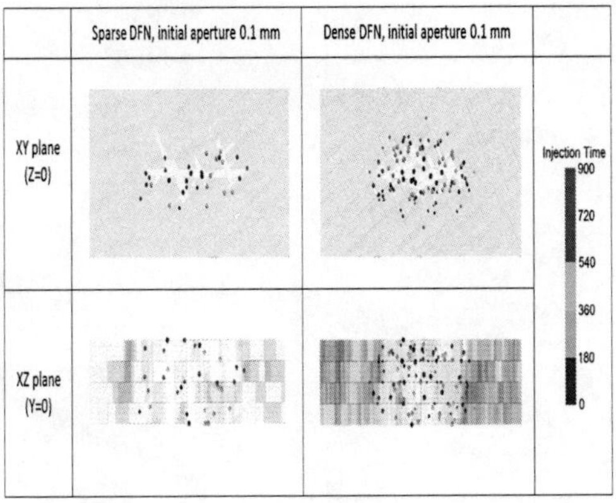

	Sparse DFN, initial aperture 0.1 mm	Dense DFN, initial aperture 0.1 mm	
XY plane (Z=0)			Injection Time 900 720 540 360 180 0
XZ plane (Y=0)			

FIGURE 4. Synthetic microseismic events for the sparse and dense DFN with the same initial aperture 0.1 mm after 900 s of injection time. The synthetic microseismic events are colored by the occurring time and sized by moment magnitude and then projected in the XY plane with z=0 m and the XZ plane with y=0 m (HF plane).

Figure 5 shows the quantitative evolution of the stimulated DFN area with a fracture pressure increase greater than 0.5 MPa and 1 MPa for the four cases shown in Figure 3. Figure 5 shows that, with the same DFN configuration, the case with a 0.15 mm initial aperture produced about twice the stimulated DFN area as the case with a 0.1 mm initial aperture. However, with the same initial aperture, the case with the dense DFN produced only slightly more stimulated DFN area than the case with a sparse DFN. Figure 5 also shows that the effect of choosing a different fracture pressure increase threshold to define the stimulated DFN areas is more obvious for the dense DFN case.

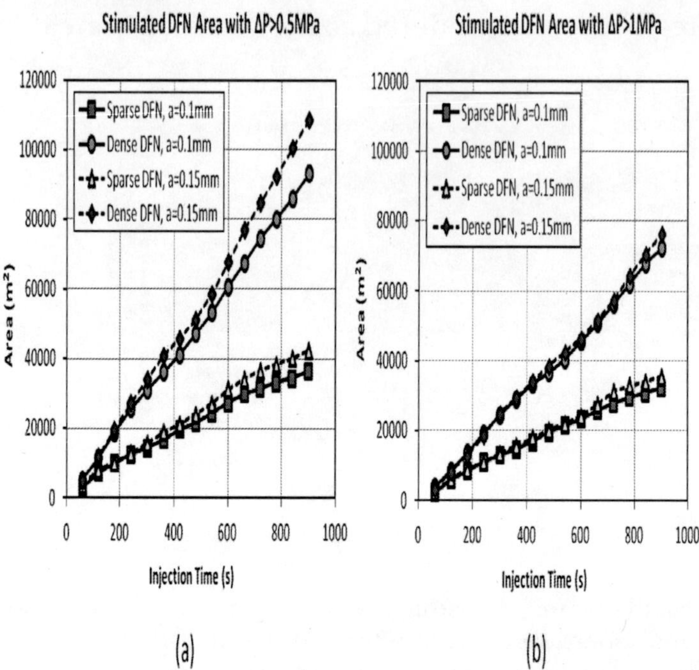

FIGURE 5. Stimulated DFN area with a fracture pressure increase greater than A) 0.5 MPa and B) 1 MPa for the four cases shown in Figure 3.

Figure 6 plots the evolution of the stimulated HF area with a fracture pressure increase greater than 0.5 MPa and 1 MPa for the four cases shown in Figure 3. Figure 6 shows that, for the sparse DFN configuration, the case with a 0.15 mm initial aperture produced only slightly less stimulated HF area than the case with a 0.1 mm initial aperture, whereas for the dense DFN configuration, there was nearly a 20% reduction in stimulated HF area for the 0.15 mm initial aperture over the 0.1 initial aperture case. Comparing the same initial aperture cases, the cases with the dense DFN produced much less stimulated HF area than the cases with a sparse DFN. Figure 6 also shows that the stimulated HF areas for all four cases using the 0.5 MPa fracture pressure increase criteria were only slightly larger than the cases when using the 1.0 MPa fracture pressure increase criteria.

It may be concluded from the above analysis that, in terms of the stimulated DFN area, the initial DFN aperture is a more sensitive parameter than the DFN density. On the contrary, in terms of the stimulated HF areas, the DFN density seems to be a more sensitive parameter than the initial DFN aperture.

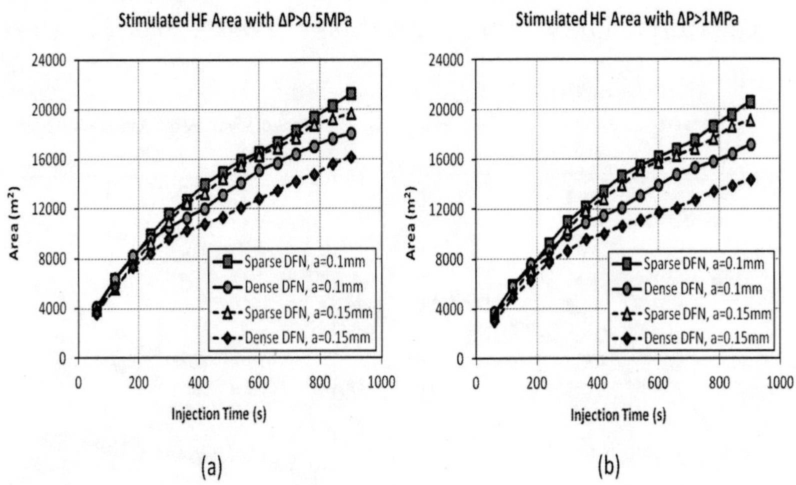

FIGURE 6. Stimulated HF area with a fracture pressure increase greater than A) 0.5 MPa and B) 1 MPa for the four cases shown in Figure 3.

The overall effect of fracture network connectivity on hydraulic fracturing effectiveness can also be characterized by the ratio of stimulated DFN area to stimulated HF area. Figure 7 plots the ratio of stimulated DFN area to stimulated HF area with a fracture pressure increase of 0.5 MPa and 1 MPa for the four cases shown in Figure 3. It can be seen that a dense DFN created a much higher ratio of stimulated DFN area to stimulated HF area than did a sparse DFN for similar initial apertures. However, the effect of initial DFN aperture was more evident for the dense DFN configuration than for the sparse DFN configuration. Meanwhile, the effect of choosing a different fracture pressure increase threshold for the stimulated areas was more obvious in the dense DFN configuration than for the sparse DFN configuration.

Another parameter to evaluate the overall effect of fracture network connectivity on hydraulic fracturing effectiveness is the leakoff ratio, which is defined as the ratio of fluid volume in the DFN over the total fluid volume injected into the model. Figure 8 shows the leakoff ratio for the four cases shown in Figure 3. Very similar with the analysis for the ratio of stimulated DFN area to stimulated HF area, DFN density is shown to significantly affect the leakoff ratio for both initial aperture cases, while the initial DFN aperture is seen to affect the dense DFN more than the sparse DFN configuration.

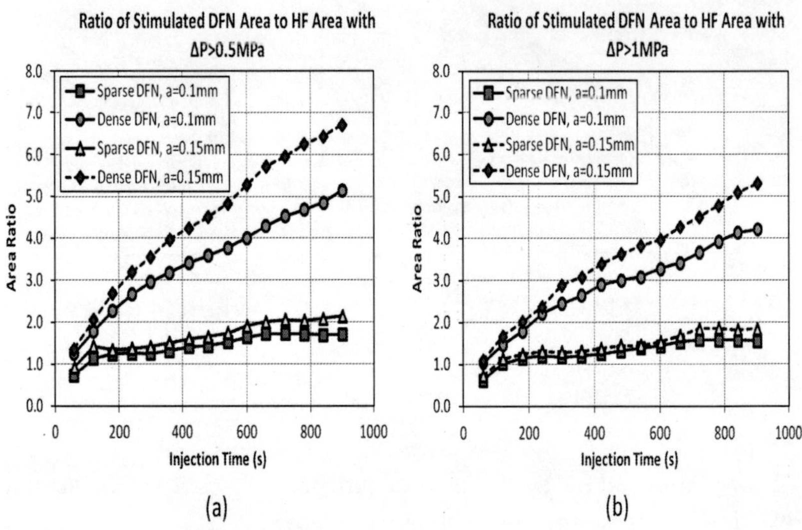

FIGURE 7. Ratio of stimulated DFN area to stimulated HF area with a fracture pressure increase greater than A) 0.5 MPa and B) 1 MPa for the four cases shown in Figure 3.

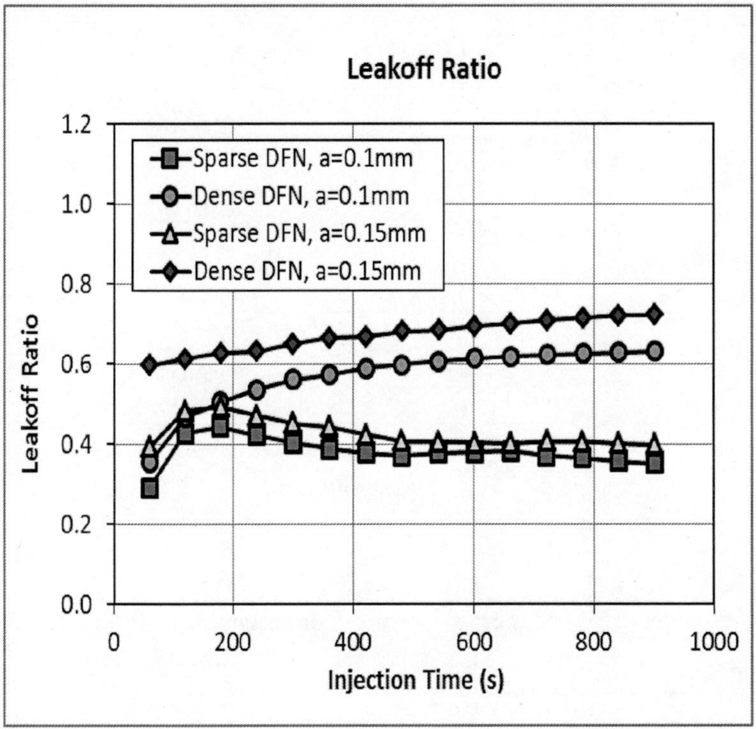

FIGURE 8. Leakoff ratio for the four cases shown in Figure 3.

Figure 9 and Figure 10 show, separately, the average DFN aperture and average HF aperture for the four cases shown in Figure 3. The average DFN aperture for all cases increased only slightly during the injection, though it is worth mentioning that even though the natural fractures were only slightly opened, the leakoff ratio for the dense DFN cases reached about 50% or more.

Relatively, the sparse DFN case showed a slightly greater increase of average DFN aperture than the dense DFN case with the same initial DFN aperture. As expected, the average HF apertures for all cases were several times larger than the average DFN apertures. In addition, as shown in Figure 9, the sparse DFN case had a much greater increase (more than double) in the average HF aperture than the dense DFN case with the same initial DFN aperture.

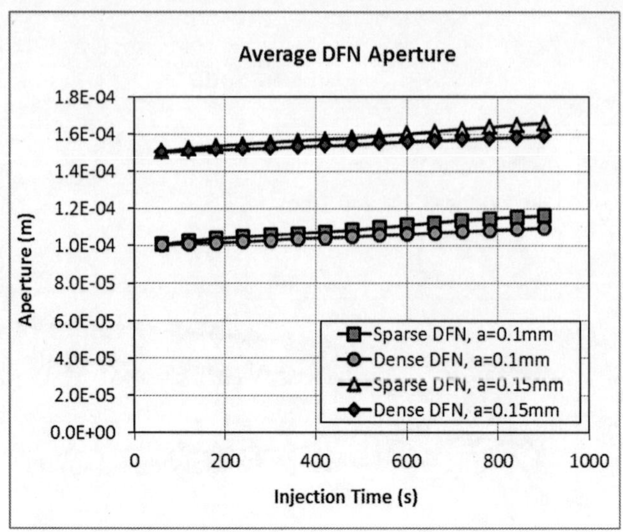

FIGURE 9. Average DFN aperture for the four cases shown in Figure 3.

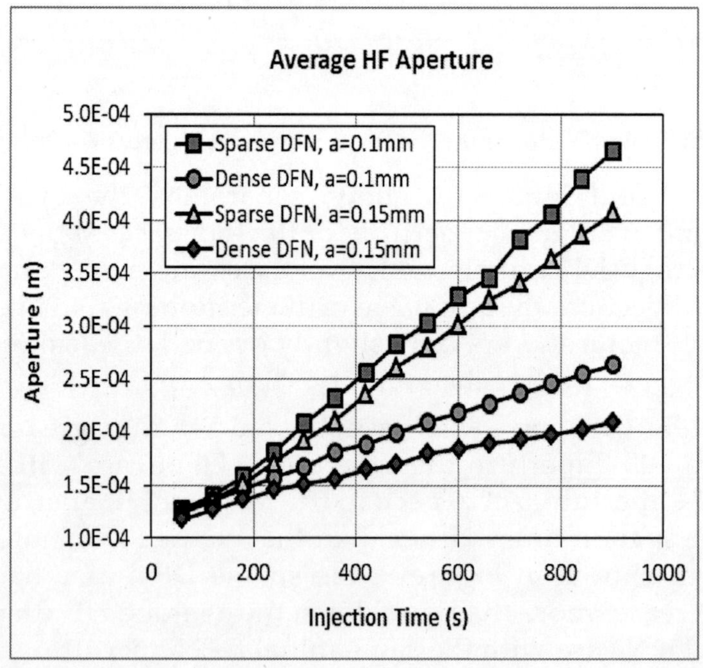

FIGURE 10. Average HF aperture for the four cases shown in Figure 3.

CONCLUSIONS

In this work, the effect of fracture network connectivity on hydraulic fracturing effectiveness (extent of stimulation of the natural fractures) was investigated using a discrete element numerical model. Four simulation cases were evaluated using two DFN configurations (i.e., a dense DFN and a sparse DFN) and two initial DFN apertures (i.e., 0.1 mm and 0.15 mm). The main conclusions from this study are summarized in the following points:

- DFN density significantly affected hydraulic fracturing effectiveness, characterized by either the ratio of stimulated DFN area to stimulated HF area or the leakoff ratio, for both initial apertures considered. Further, the initial DFN aperture affected the hydraulic fracturing effectiveness of the dense DFN configuration more than the sparse DFN configuration.
- The sparse DFN cases showed a flat microseismic distribution zone with few events while the dense DFN cases showed a complex microseismic map that indicated the intensive interaction between hydraulic fracture and the natural fractures.
- For all cases, the average DFN aperture increased only slightly during the injection while the average HF aperture increased significantly and was several times larger than the average DFN apertures. Relatively, the sparse DFN case showed a greater increase of average DFN aperture and average HF aperture than the dense DFN case for the same initial DFN aperture.
- This work suggests that fracture network connectivity plays a critical role in hydraulic fracturing effectiveness for unconventional shale developments, and fracture connectivity will play a significant role in optimizing treating pressures, the created microseismicity and corresponding SRV, and well production.

REFERENCES

1. G. E. King, "Thirty years of gas shale fracturing: What have we learned?," in SPE Annual Technical Conference and Exhibition, Florence, Italy, 2010.
2. C. Cipolla, M. Mack and S. Maxwell, "Reducing Exploration and

Appraisal Risk in Low-Permeability Reservoirs Using Microseismic Fracture Mapping," in Canadian Unconventional Resources and International Petroleum Conference, Calgary, Alberta, Canada, 2010.

3. N. R. Warpinski, "Integrating Microseismic Monitoring With Well Completions, Reservoir Behavior, and Rock Mechanics," in SPE Tight Gas Completions Conference, San Antonio, Texas, USA, 2009.
4. S. F. Rogers, D. Elmo and W. S. Dershowitz, "Understanding Hydraulic Fracturing Geometry and Interactions in Pre-Conditioning through DFN Numerical Modeling," in 45th US Rock Mechanics / Geomechanics Symposium, San Francisco, CA, 2011.
5. W. S. Dershowitz, M. G. Cottrell, D. H. Lim and T. W. Doe, "A discrete fracture network approach for evaluation of hydraulic fracturing stimulation of naturally fractured reservoirs," in 44th US Rock Mechanics Symposium and 5th U.S.-Canada Rock Mechanics Symposium, Salt Lake City, 2010.
6. N. B. Nagel, M. A. Sanchez-Nagel, F. Zhang, X. Garcia and B. Lee, "Coupled Numerical Evaluations of the Geomechanical Interactions Between a Hydraulic Fracture Stimulation and a Natural Fracture System in Shale Formations," Rock Mechanics and Rock Engineering, 2013.
7. N. B. Nagel, X. Garcia, M. A. Sanchez-Nagel and B. Lee, "Understanding "SRV": A Numerical Investigation of "Wet" vs. "Dry" Microseismicity During Hydraulic Fracturing," in SPE Annual Technical Conference and Exhibition, San Antonio, Texas, USA, 2012.
8. N. B. Nagel, M. A. Sanchez-Nagel, X. Garcia and B. Lee, "A Numerical Evaluation of the Geomechanical Interactions Between a Hydraulic Fracture Stimulation and a Natural Fracture System," in 46th US Rock Mechanics / Geomechanics Symposium, Chicago, IL, 2012.
9. N. Nagel, I. Gil, M. Sanchez-Nagel and B. Damjanac, "Simulating hydraulic fracturing in real fractured rock - overcoming the limits of pdeudo-3D models," in SPE 140480 presented at the SPE Hydraulic Fracturing Technology Conference, the Woodlands, TX, 2011.
10. A. A. Savitski, M. Lin, A. Riahi, B. Damjanac and B. N. Nagel, "Explicit Modeling of Hydraulic Fracture Propagation in Fractured Shales," in International Petroleum Technology Conference, Beijing, China, 2013.
11. ICG, 3DEC - Three-Dimensional Distinct Element Code, Version 4.2, Minneapolis, MN: Itasca Consulting Group, Inc., 2007.

Chapter 3

THE IMPACT OF THE NEAR-TIP LOGIC ON THE ACCURACY AND CONVERGENCE RATE OF HYDRAULIC FRACTURE SIMULATORS COMPARED TO REFERENCE SOLUTIONS

Brice Lecampion, Anthony Peirce, Emmanuel Detournay, Xi Zhang, Zuorong Chen, Andrew Bunger, Christine Detournay, John Napier, Safdar Abbas, Dmitry Garagash and Peter Cundall

ABSTRACT

We benchmark a series of simulators against available reference solutions for propagating plane-strain and radial hydraulic fractures. In particular, we focus on the accuracy and convergence of the numerical solutions in the important practical case of viscosity dominated propagation. The simulators are based on different propagation criteria: linear elastic fracture mechanics (LEFM), cohesive zone models/tensile strength criteria, and algorithms accounting for the multi-scale nature of hydraulic fracture propagation in the near-tip region. All the simulators tested here are able to capture the analytical solutions of the different configurations tested, but at vastly different computational costs. Algorithms based on the classical LEFM propagation condition require a fine mesh in order to capture viscosity dominated hydraulic fracture evolution. Cohesive zone models, which model the fracture process zone, require even finer meshes to obtain the same accuracy. By contrast, when the algorithms use the appropriate multi-scale hydraulic fracture asymptote in the near-tip region, the exact solution can be matched accurately with a very coarse mesh. The different analytical reference solutions used in this paper provide a crucial series of benchmark tests that any successful hydraulic fracturing simulator should pass.

INTRODUCTION

The propagation of hydraulic fractures is a highly non-linear fluid-solid interaction problem involving a moving boundary (i.e., the propagating fracture front in the neighboroughood of which the governing equations degenerate). Simulating this class of problem numerically is challenging, especially properly tracking the evolving fracture front.

Table 1. List of available solutions for the propagation of hydraulic fractures driven by a Newtonian fluid under constant rate of injection (zero leak-off).

Propagation regimes	Plane-strain	Radial
Viscosity M ($\mathcal{K} = 0$)	[3] (with correction for small toughness)	[6]
Toughness K ($\mathcal{K} \to \infty$)	[4] (with correction for small viscosity)	[6, 11]

In geoscience applications, hydraulic fractures propagate in a complex, often poorly characterized medium. Nevertheless, the description of the medium must be simplified in order to apply theoretical models. It is thus crucial that numerical implementations of such models for fracture growth be accurate such that differences from field observations can be attributed to model assumptions rather than poor numerical solution.

In the last ten years, a number of reference solutions (analytical and semi-analytical) have been obtained for propagating plane-strain [1–5] and radial hydraulic fractures [6, 7] (see Table 1). These solutions provide invaluable benchmarks for numerical simulators. We compare a number of simulators (2D and 3D) that use different propagation algorithms against these reference solutions for hydraulic fractures driven by a Newtonian fluid under a constant injection rate. For the sake of clarity, we do not address fluid leak-off in our discussion. Of particular interest is the accuracy of the different simulators in tracking the moving fracture front, particularly in the so-called viscosity-dominated regime of propagation.

An outstanding question relates to the convergence and robustness of numerical simulators with respect to the multiscale near-tip behavior of hydraulic fractures. The coupled lubrication (fluid flow) and elasticity equations are known to degenerate near the fracture tip, such that the solution of a semi-infinite fracture propagating at a constant velocity is characterized by a multiscale singular behavior near the tip [8–10]. The nature of the dominant singularity depends on the relative importance of two dissipative processes (viscous forces and fracture energy), as well as the reference length scale. Such a multiscale behavior near the fracture tip in turn governs the evolution of the velocity of a finite hydraulic fracture during injection. We discuss the degree to which a numerical simulator needs to include and resolve the near-tip behavior in order to accurately match reference solutions in the light of different benchmarks.

BENCHMARKS

Plane-strain hydraulic fracture (KGD)

The case of a plane strain hydraulic fracture driven by a Newtonian fluid under constant injection rate is also sometimes refered to as the KGD model (for Khristianovic [12],

Geerstma and De klerk [13]). In the absence of leak-off the solution of the hydraulic fracture propagation is self-similar and depends on a single dimensionless number: i.e. a dimensionless toughness K

$$\mathcal{K} = \frac{K'}{E'^{3/4}\mu'^{1/4}Q_o^{1/4}}$$

Where Q_o is the volumetric injection rate per unit length in the out-of plane direction, E ' denotes the plane-strain elastic modulus, μ ' = 12μf is an equivalent viscosity (with μf the fluid viscosity), and K ' = √ 32/π K_{Ic} with K_{Ic} the fracture toughness (see [14] for more details). Equivalently a dimensionless viscosity M = K−4 can be used. The complete solutions for the fracture length evolution, fracture width and net pressure have been obtained for the limiting cases of zero dimensionless toughness (equivalently infinite M) and zero dimensionless viscosity (infinite K) First order solutions for either small toughness or viscosity are also available [3, 4]. Semi-analytical solutions for any finite values of dimensionless toughness or viscosity are also available [5].

We will restrict our comparisons to the case of relatively small dimensionless toughness (e.g. K < 1), which is known to be the more difficult condition to reproduce numerically. Therefore we express the solution in a so-called viscosity scaling. Because the solution is self-similar the time dependence can be obtained using dimensional analysis. We aim to compare the solutions provided by different numerical codes, which are typically developed in space-time. We thus introduce a dimensionless time $\tau = t/t_c$ and a scaled coordinate $\xi = x/\ell$, where t_c is a characteristic time scale and ℓ is the fracture length. The fracture length, opening, and net pressure can be written as follows:

$$\ell = \left(\frac{E'Q_o^3 t^4}{\mu'}\right)^{1/6} \gamma_m(\mathcal{K}) = \left(\frac{E'Q_o^3 t_c^4}{\mu'}\right)^{1/6} \underbrace{\tau^{2/3}\gamma_m(\mathcal{K})}_{\gamma(\tau,\mathcal{K})}$$

$$w = \frac{Q_o^{1/2} t^{1/3} \mu'^{1/6}}{E'^{1/6}} \Omega_m(\xi,\mathcal{K}) = \frac{Q_o^{1/2} t_c^{1/3} \mu'^{1/6}}{E'^{1/6}} \underbrace{\tau^{1/3}\Omega_m(\xi,\mathcal{K})}_{\Omega(\xi,\tau,\mathcal{K})} \tag{1}$$

$$p = \frac{E'^{2/3}\mu'^{2/3}}{t_c^{1/3}} \Pi_m(\xi,\mathcal{K}) = \frac{E'^{2/3}\mu'^{2/3}}{t^{1/3}} \underbrace{\tau^{-1/3}\Pi_m(\xi,\mathcal{K})}_{\Pi(\xi,\tau,\mathcal{K})}$$

where we have highlighted the correspondence between results obtained using a time-based algorithm (say γ, Ω, Π) to the self-similar solution dimensionless solution $F_m(K) = \{\gamma_m, \Omega_m, \Pi_m\}$.

The dimensionless solution $F_m(K) = \{\gamma_m, \Omega_m, \Pi_m\}$ for small toughness developed in [3] will be compared with the numerical solutions from different simulators. More precisely, for three small values of dimensionless toughness K = 0.01, 0.1, 0.5, we will focus on the comparisons of the dimensionless fracture length γ_m, opening profile $\Omega_m(\xi)$ close to the fracture tip and error in the fracture volume etc. We are especially interested in the evolution of the error of the numerical solutions with respect to mesh sizes in the near-tip region of the fracture, where gradients are the largest. The solution, for small toughness (K < 1), is in fact governed by the hydraulic fracture viscosity tip asymptote: the opening behaves as w ~ (ℓ − x) 2/3 and the net pressure as p ~ (ℓ − x) −1/3 close to the fracture tip (see [10] for details). The tip region affected by the asymptote actually extends to about 10 to 20 percent of the plane-strain fracture length for dimensionless toughness's below 0.5.

RADIAL HYDRAULIC FRACTURE

The growth of a radial hydraulic fracture spans both the viscosity and toughness regimes of propagation [6, 11]. At early times, the perimeter and the opening of the fracture are small and most of the energy is spent in viscous flow, whereas at a later times, the fracture perimeter and opening are larger and the fracture energy required to extend the fracture dominates the energy required to drive the viscous fluid through the fracture. The radial solution is also dependent only on a dimensionless toughness which in this case is a function of time. Introducing the characteristic time

$$t_{mk} = \left(\frac{\mu'^5 Q_o^3 E'^{13}}{K'^{18}} \right)^{1/2},$$

, and the dimensionless time $\tau = t/t_{mk}$ we have (see [6, 14] for more details):

$$\mathcal{K} = \tau^{1/9}$$

Solutions for the case of zero and infinite dimensionless toughness (i.e. small and large dimensionless time) have been obtained semi-analytically [6]. The complete transient solution can be obtained only numerically. A reference algorithm [7] based on an explicit moving mesh algorithm with proper matching of the multiscale HF tip asymptotics [10] will provide the baseline for the comparisons for intermediate times. The fracture radius R, width ω and net pressure p can be written as:

$$\begin{aligned} R &= \frac{E'^3 Q_o \mu'}{K'^4} \gamma(\tau) \\ w &= \frac{\sqrt{E' Q_o \mu'}}{K'} \Omega(\rho, \tau) \\ p &= \frac{K'^3}{E'^{3/2} Q_o^{1/2} \mu'^{1/2}} \Pi(\rho, \tau) \end{aligned} \tag{2}$$

where the dimensionless solution $F = \{\gamma, \Omega, \Pi\}$ depends only on dimensionless time $\tau = t/t_{mk}$ and scaled position $\rho = r/R$ along the fracture. As before, we will pay particular attention to the case of small dimensionless toughness, i.e. early-time, which is the most challenging numerically. In the limit of zero-toughness/early-time (i.e. viscosity dominated propagation, here refereed as the M-vertex), the solution is self-similar and can be conveniently written in the following viscosity scaling:

$$
\begin{aligned}
R &= \frac{E'^{1/9} Q_o^{1/3} t^{4/9}}{\mu'^{1/9}} \gamma_{m0} = \frac{E'^3 Q_o \mu'}{K'^4} \underbrace{\tau^{4/9} \gamma_{m0}}_{\gamma(\tau)} \\
w &= \frac{Q_o^{1/3} \mu'^{2/9} t^{1/9}}{E'^{2/9}} \Omega_{m0}(\rho) = \frac{\sqrt{E' Q_o \mu'}}{K'} \underbrace{\tau^{1/9} \Omega_{m0}(\rho)}_{\Omega(\rho,\tau)} \qquad (3) \\
p &= \frac{E'^{2/3} \mu'^{1/3}}{t^{1/3}} \Pi_{m0}(\rho) = \frac{K'^3}{E'^{3/2} \sqrt{Q_o \mu'}} \underbrace{\tau^{-1/3} \Pi_{m0}(\rho)}_{\Pi(\rho,\tau)}
\end{aligned}
$$

where again, we have highlighted the correspondence between the zero-toughness/ M-vertex self-similar solution Fm0 = {γm0,Ωm0,Πm0}, which is independent of time and the dimensionless solution expressed as a function of the previously defined dimensionless time.

The zero-toughness/M-vertex solution Fm0 has actually been found to correctly capture the propagation of hydraulic fractures up to a dimensionless toughness of K = 1, i.e. for dimensionless time $\tau \leq 1 (t \leq t_{mk})$ [6]. We will thus investigate the convergence of different simulators to this zero-toughness/small time solution.

For dimensionless time above unity ($\tau > 1 (t > t_{mk})$), the solution transitions from the viscosity dominated (early-time) to the toughness (large-time) dominated regime. The toughness dominated regime is reached for $K \approx 3.5 (\tau \approx 70000)$. Note that, for infinitely large dimensionless toughness (i.e. zero viscosity), the solution is also self-similar [6] and is also denoted as the K-vertex solution. We will also briefly investigate, for a subset of the simulators considered, the transition between these two regimes of propagation (M to K), focusing mostly on fracture length versus time.

SOME PRACTICAL NUMBERS

Although rock properties and stimulation practices vary, it is interesting to compute the scales and dimensionless numbers previously introduced. Let's assume a "tight" rock with the following realistic properties: a plane-strain Young's modulus of 40 GPa and a fracture toughness of 1.5 MPa.m1/2. First, for a hydraulic fracturing treatment using a highly viscous fluid (e.g. gel-like with μf = 100 cPoise) at a practical rate of 10 Barrels per minute, we obtain a transition time scale t_{mk} for a radial fracture of 4.2 106 seconds! The propagation of such a hydraulic fracture for a realistic injection duration (i.e. less than two hours) will always be in the viscosity dominated regime of propagation. Remember that the dimensionless toughness evolves as (t/t_{mk}) 1/9 for radial fractures. For the plane-strain geometry (where the injection rate is per meter in the out-of-plane direction), using the same parameters we obtain a dimensionless toughness of 0.26 clearly indicating a viscosity dominated propagation.

For a slick-water treatment, popular in shale-gas reservoirs, the injection rates are usually much higher in order to compensate for the low viscosity of water and to obtain a sufficiently wide fracture to accommodate prop pant (see the scales in front of the opening ω in Equations (1) and (3)). Using a value of 20 Barrels per minute (a realistic value for a single perforation cluster) and the viscosity of water, the radial transition time scale t_{mk} now reduces to two minutes. For radial fractures, the toughness dominated regime of propagation is obtained for dimensionless toughness above 3.5 (see [6]), which corresponds to t & 75000tmk, which translate for this particulate case to t & 250 hours. This indicates that most of the duration of the treatment will take place in the transition from the viscosity to the toughness regimes of propagation for a radial fracture (assuming that no stress barriers affect the fracture geometry). For the plane-strain fracture geometry, we obtain a dimensionless toughness of 0.3 (assuming the same injection rate per meter in the out-of-plane direction), for which the propagation is still dominated by viscosity.

These examples show the importance of the viscosity dominated regime of propagation for oil and gas hydraulic fracturing applications. Numerical simulators therefore need to be able to capture this regime of propagation, which is a difficult task especially if the algorithm relies solely on the linear elastic fracture mechanics propagation condition that manifests itself at a length scale near the fracture tip that is much smaller than the modeling length scale in the viscosity dominated case [10, 15].

SIMULATORS TESTED

Two classes of simulators have been tested: codes simulating a two-dimensional configuration (plane-strain or/and axisymmetry) where the fracture is a one dimensional geometrical object, and three dimensional codes simulating planar fractures (which are two-dimensional objects in three dimensions). We now briefly describe the algorithms used by these different simulators.

TWO DIMENSIONAL CODES

MineHF2D

This simulator (see [16] for more details) handles the propagation of both straight and curved hydraulic fractures in plane-strain. The algorithm is based on a fixed grid. It uses the displacement discontinuity method to solve the elastic equations coupled to a finite difference scheme for the fluid flow within the fracture. This algorithm includes the presence of a fluid lag at the fracture tip; for the simulated case reported here, a large confining stress value was used to minimize the fluid lag (see [17, 18] for more discussion on the effect of fluid lag). Explicit time-stepping is used and a volume of fluid method locates the fluid front. The fracture propagation criterion is based on the linear elastic fracture mechanics asymptote. The stress intensity factors are obtained using the displacement method with an adjusting factor of 0.88. A mesh with variable element sizes was used with refinement toward the fracture tip.

FEM_Cohesive

This code is based on a finite element model and the pore pressure cohesive element implemented in Abaqus (Abaqus 6-10.2, 2010). It can handle both plane-strain and axi-symmetric configurations (e.g. plane-strain and radial hydraulic fractures). In this model, a pre-defined surface made up of elements that support the cohesive zone traction-separation calculation is embedded in the rock and the hydraulic fracture grows along this pre-defined surface. The fracture process zone (unbroken cohesive zone) is defined within the separating surfaces where the surface tractions are nonzero. The fracture is fully filled with fluid in the fully damaged cohesive zone (where the cohesive traction is zero) and hence there will be no cohesive traction contribution, but fluid pressure is acting on the open fracture surfaces. So a coupled fluid pressure-traction-separation relationship exists between the cohesive zone defined by the traction-separation law and the pressurized fracture as found from solving the lubrication equation with the constraint that all tractions acting on the entire fracture and the cohesive zone must be in equilibrium. In this cohesive finite element model [19], the irreversible bilinear traction-separation cohesive law is adopted. An incompressible Newtonian fluid is injected at the center of the fracture at constant injection rate. There is no fluid leak-off through the impermeable surfaces of the fracture, so only flow in the fracture radius direction is modelled. The cohesive elements at the injection point are defined as initially open to allow entry of fluid, and so that the initial flow and fracture growth is possible. Infinite elements surrounding the finite domain, which contains a hydraulic fracture, have been used to model the far-field boundary. Further details of the finite element model can be found in [19].3. 1DPlanarHF

This code, also based on a fixed mesh, simulates straight hydraulic fractures in two-dimensions (plane-strain and axisymmetric fractures) using a fully implicit scheme to solve for the coupling between the elasticity equation (discretized using the displacement discontinuity method), the fluid conservation (discretized using a finite volume scheme) and to locate the fracture front. An increment of fracture length is given and the corresponding time-step (to reach the new fracture length) is solved by satisfying the fracture propagation condition in the tip element in a weak form: i.e. the volume of the tip element is enforced to be equal to the LEFM square-root asympote (The algorithm is similar to the one described in [20], see also [21]). The HF viscosity tip asymptote [8] can also be used for the case of low fracture toughness, its performance will be compared to the LEFM asymptote. Results obtained using the LEFM and the viscosity asymptotes will be denoted as 1DPlanarHF_lefm, and 1DPlanarHF_m respectively. All the results presented here use a grid with a constant element size (i.e. without any refinements), a re-coarsening of the mesh during the simulation is possible.4. EMMA

EMMA is an Explicit Moving Mesh Algorithm for radial geometry, which embeds the proper multiscale tip asymptotes of the hydraulic fracture depending on its velocity (see [7] for more details). It is extremely accurate and the moving mesh nature of the algorithm allows it to span more than ten orders of magnitude in dimensionless time. It notably provides a good solution for the transition between the viscosity and toughness dominated regime of propagation for a radial fracture.

THREE DIMENSIONAL CODES

The Implicit Level Set Algorithm (ILSA)

This algorithm [22] models the evolution a hydraulic fracture with an arbitrarily shaped boundary that is assumed to propagate in a plane (a planar fracture in a 3D elastic medium), which is typically perpendicular to the minimum principal stress direction. The three dimensional elastic equilibrium equations are discretized using the displacement discontinuity boundary integral method in which the fracture within the plane is represented by constant width rectangular elements that are collocated at element centres.The Reynolds lubrication equation, expressing the conservations of mass of the viscous fluid contained within the crack surfaces, is discretized using a finite volume method also defined with respect to quantities sampled at the centres of the rectangular elements. At the periphery of the fracture, which may not conform to the structured rectangular mesh, the boundary is represented using a concept of partially filled tip elements that are used to define average fracture widths, which are also sampled at element centres. The distinguishing feature of this algorithm is its ability to locate the fracture free boundary using the asymptotic behavior of the hydraulic fracture width that is applicable at a particular point on the fracture perimeter. The free boundary is located by the following iterative process: given an initial guess for the fracture boundary ∂S, determine the corresponding trial fracture width w and fluid pressure field pf ; in the ribbon of elements that are completely filled with fluid and, which share at least one side with a partially filled tip element, use the trial width values to estimate the distance to the free boundary by inverting the applicable tip asymptotic behavior [10]; use these estimates of the distance to the free boundary as initial conditions for the eikonal equation $|\nabla T(x,y)| = 1$, whose level set curve $T(x,y) = 0$ is the free boundary. The fracture boundary is then moved to the curve $T(x,y) = 0$ and the iterative process is repeated until convergence is achieved. The algorithm uses the multi-scale hydraulic fracture tip asymptotics solution [10] and thus automatically captures the different type of propagation regimes with relatively coarse mesh.

For this paper, a simplified version of the algorithm was also designed to only use the LEFM asymptote (hereafter denoted as ILSA_lefm) for comparisons with other algorithms (MineHF2D, 1DPlanarHF etc.). In this version, we adapted the ILSA code to damp the front advance by rescaling the level set function $T(x,y)$ so that the the maximum distance between any point in the ribbon elements and the damped free boundary is no more than three element lengths. This sequence of damped front positions enables the trial widths to be relaxed until fracture width profile presents a close approximation to the viscosity dominated solution, in spite of the fact that the tip elements are, by the nature of the ILSA_lefm algorithm, locked into the LEFM asymptote.

HFLattice

This code [23] simulates fracture propagation without limitation of shape, direction or number of fractures, as well as slip and opening along pre-existing joints. A 3D lattice formulation is used for simulation of deformation and fracturing. The lattice is a quasi-random assembly of nodes connected by non-linear shear and normal springs. The lattice resolution is given by the average node spacing. Newton's law of motion (for translation and rotation) is solved at the nodes using an explicit central difference scheme. The normal force in the spring is tested and a micro-crack is formed when breakage is detected (spring strength is adjusted to give the correct rock strength). A macro-fracture that develops in intact rock is thus characterized by an assembly of micro-cracks. Fluid flow and storage are based on a network of fluid nodes, located at broken springs or springs intersected by pre-existing joints, connected by pipes. The fluid network is updated continuously as new fracturing occurs. An explicit fluid pressure scheme is used to solve for fracture and matrix flow. The mechanical and flow models are fully coupled:

fracture permeability depends on aperture (i.e. deformation of the mechanical components), fluid pressure affects deformation and strength of the solid model, and deformation of the solid model affects fluid pressures. In the algorithm, the lattice springs carry total forces, which affects force balance and motion. Also, effective stress is considered for joint slip or opening.

Table 2. The benchmarks tested (X) for the different simulators.

	Plane-strain		Radial	
	$\mathcal{K} = 0.01$	$\mathcal{K} = 0.1$	$\tau \ll 1$ ($\mathcal{K} \ll 1$)	$10^{-1} < \tau < 10^4$ ($.5 < \mathcal{K} < 3.5$)
MineHF2D	✓	✓	n.a	n.a
FEM_Cohesive	✓	✓	✓	
1DPlanarHF	✓	✓	✓(_lefm only)	✓(_lefm only)
ILSA	n.a	n.a	✓	✓
HFLattice	n.a	n.a	✓	

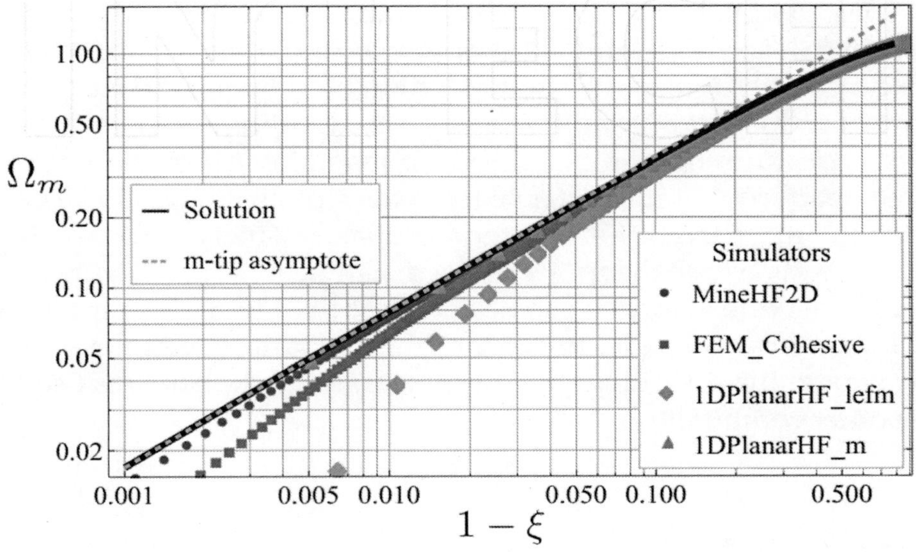

Figure 1. Dimensionless opening Ωm from the fracture tip in log-log scale; plane-strain fracture K = 0.01.

RESULTS AND DISCUSSION

The plane strain small toughness benchmark

The solution for a plane-strain hydraulic fracture driven by the injection of a Newtonian fluid at a constant rate is self-similar (i.e. evolves as a power-law of time). Our comparisons here focus on the case of viscosity dominated fractures (K < 0.5, especially K = 0.1, 0.001) which are the most difficult to simulate numerically. The simulators tested for that configuration are (Table 2): MineHF2D, FEM_Cohesive, 1DPlanarHF_lefm and 1DPlanarHF_m.

Figure 1 displays, for the case K = 0.01 the dimensionless fracture opening profiles from the tip of the fracture obtained with the different simulators (at the last step of their simulations) as well as both the analytical solution and the viscosity HF tip asymptote [10]. Figure 2 is similar but for the case K = 0.1. We first observe that the viscosity tip asymptote covers a region about 10 to 20 percent of the fracture from its tip. The different simulators provide width estimates that all correctly fall on the analytical solution "away" from the fracture tip. The distances from the tip at which the simulators recover the analytical solution appear to depend on both the mesh-size and the type of propagation condition used. For algorithms using the linear elastic fracture mechanics (LEFM) propagation condition (opening as a square root of the distance from the tip), this recovery distance from the tip is larger for coarser mesh sizes. The algorithm using a cohesive zone model appears to need significantly more refinement. In contrast, the algorithm using the viscosity HF tip asymptote (i.e. 1DPlanarHF_m) is able to capture the fracture opening exactly all the way to the tip and with a much coarser mesh than was used for the other computations.

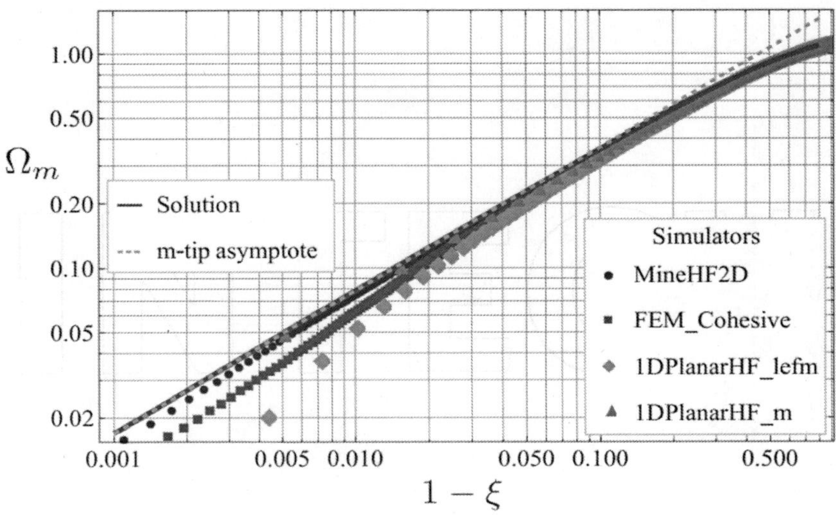

Figure 2. Dimensionless opening Ωm from the fracture tip in log-log scale; plane-strain fracture K = 0.1.

This dependence of the convergence toward the exact solution on the mesh size and propagation condition can be further observed in Figures 3-5, which display the rate of convergence for the fracture length and fracture volume as a function of the ratio of the mesh size over fracture length (i.e. the inverse of the number of elements to discretize the fracture for uniform mesh). All simulators converge correctly toward the analytical solution but at very different computational costs. We can see that for algorithms using the LEFM condition or a cohesive zone model, the mesh size required to reach the same level of accuracy is about 20 times smaller than for the algorithm that uses the correct HF viscosity tip asymptote. In the case K = 0.01, 1DPlanarHF_lefm needs about 400 elements (h/ ~ .0025) to obtain a relative error of about 1 and 3 percent in the fracture length and fracture volume respectively, while smaller relative errors are already obtained when using 20 elements (h/ ~ .05) for 1DPlanarHF_m. The cost is even greater when a cohesive zone model is used: about 3000 elements (h/ ~ 3×10−4) are needed to reach a relative error of 1.5 and three percent in the fracture length and fracture volume respectively.

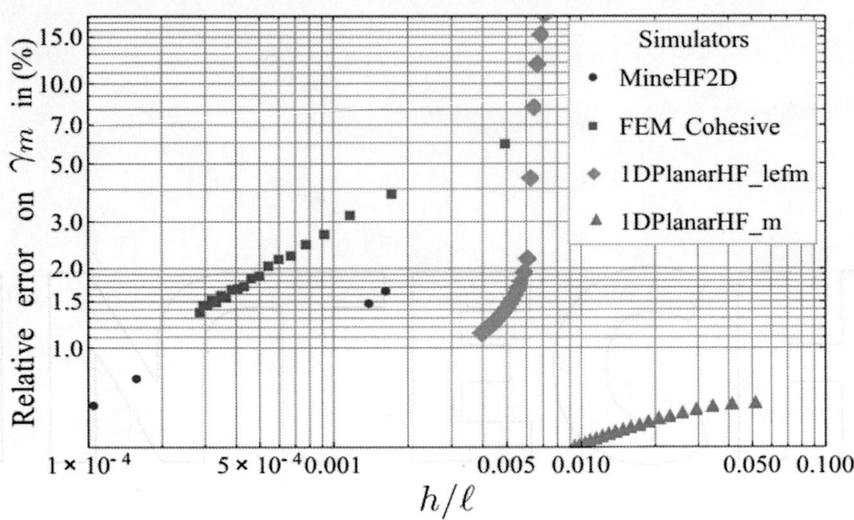

Figure 3. Relative error in the fracture length as a function of the ratio mesh-size over fracture length K = 0.01.

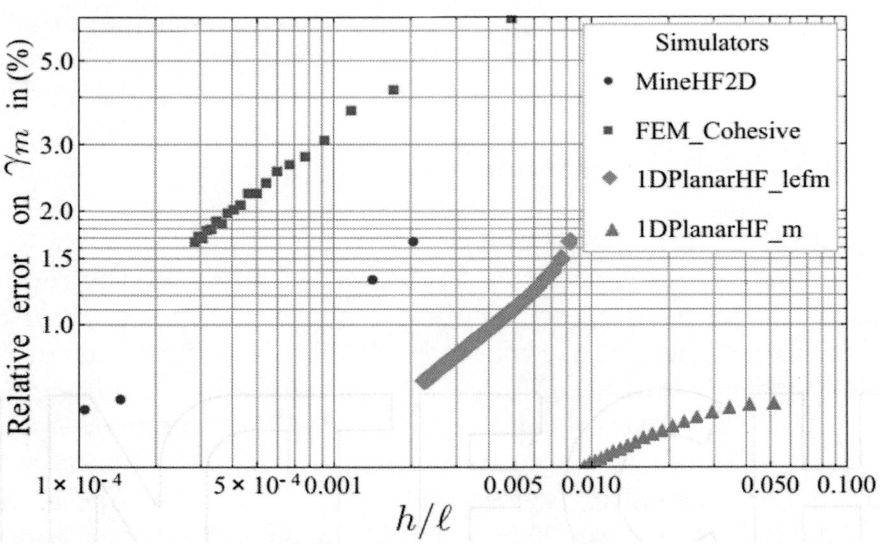

Figure 4. Relative error in the fracture length as a function of the ratio mesh-size over fracture length K = 0.1 Similar observations can be made for the case of a dimensionless toughness K = 0.1, although all the algorithms using a frac-

ture energy propagation condition perform slightly better due to the higher value of dimensionless toughness. In other words, similar relative errors are obtained for larger values of h/ℓ (i.e. fewer elements), as can be seen by comparing Figures 3 and 5 for K = 0.01 to Figures 4 and 6 for K = 0.1.

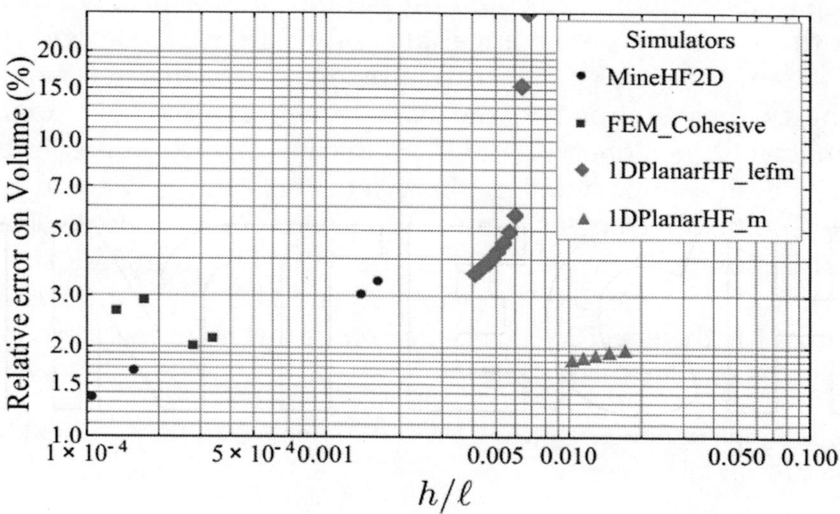

Figure 5. Relative error in the dimensionless fracture volume as a function of mesh size over fracture length - Plane-strain fracture K = 0.01.

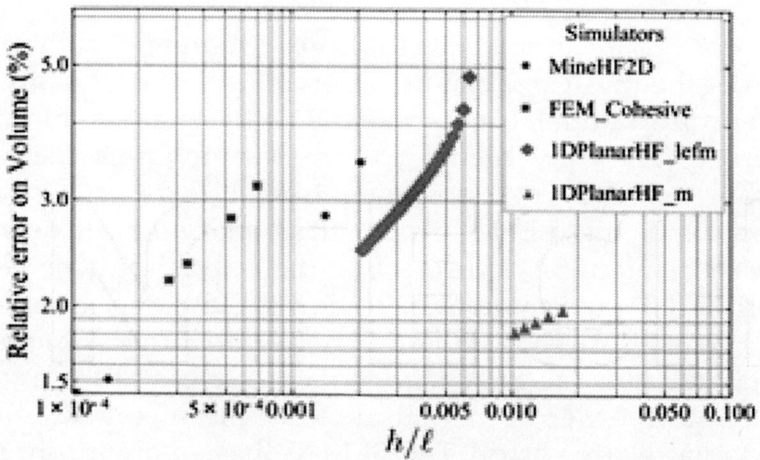

Figure 6. Relative error in the dimensionless fracture volume as a function of mesh size over fracture length - Plane-strain fracture K = 0.1.

THE RADIAL HYDRAULIC FRACTURE BENCHMARK

Viscosity dominated regime

Owing to its previously argued importance in practice, we focus on early-time, where the relevant analytical solution corresponds to the case of zero-toughness. The simulators tested for this geometry and regime of propagation are the two-dimensional codes under axi-symmetry; FEM_Cohesive and 1DPlanarHF_lefm, and the 3D codes ILSA, the simplified version ILSA_lefm and the HFLattice model (Table 2). The different simulators have been run for different ranges of dimensionless time all within the viscosity dominated regime ($\tau < 1$), and are described in Table 3. The HFLattice simulator has been run for the specific case of zero toughness (in fact zero tensile strength as per the formulation).

Table 3. Range of dimensionless time of the simulations for the radial benchmark

FEM_Cohesive	1DPlanarHF_lefm	ILSA/ILSA_lefm
$\tau \in [10^{-10} - 2 \times 10^{-8}]$	$\tau \in [10^{-7} - 10^{-1}]$	$\tau \in [10^{-18} - 10^{-17}]$

The convergence toward the zero-toughness solution as a function of the mesh size can be observed in Figure 7 (convergence of the fracture radius). Similar trends to the plane-strain case can be observed. The convergence requires a much finer mesh for the simulators using a cohesive zone model (axi-symmetric FEM_Cohesive) and the LEFM propagation condition (axisymmetric 1DPlanarHF and ILSA_lefm). ILSA, the only simulator using the appropriate HF tip asymptote, achieves the same accuracy with a much coarser mesh compared to all the other simulators. In particular, the version ILSA_lefm, which uses the LEFM asymptote, needs about an order of magnitude finer mesh compared to ILSA for the same relative error. The FEM_Cohesive algorithm requires a ratio h/R about two to three orders of magnitude smaller than ILSA for the same relative error. The HFLattice model, although less accurate, also exhibits convergence as h/R decreases.

The openings at the last time step of the simulation (refer to Table 3 for the corresponding dimensionless time) in the tip coordinate system are compared to the zero-toughness analytical solution in Figure 8. The FEM_Cohesive algorithm captures the solution away from the fracture tip well, i.e at distance larger than 3-5 % of the fracture radius from the tip. Closer to the tip, the opening from the cohesive zone algorithm appears to slightly overestimate the fracture opening. The results of the algorithms using the LEFM propagation condition (1DPlanarHF_lefm, ILSA_lefm) converge toward the analytical opening as their mesh gets finer. On the other hand, ILSA exactly matches the analytical solution for the opening with a relatively coarse mesh (the last element of ILSA corresponds to a partially fractured element for which the fracture width is also function of the fracture front location within the element).

A comparison of the net pressure profile obtained by the different simulators and the analytical solution is displayed on Figure 9. Similar trends to the fracture opening can be observed. One has to note that the HFLattice simulator approximates the point-source by a finite volume, hence the discrepancy on the net pressure with respect to the point-source solution close to the fracture inlet.

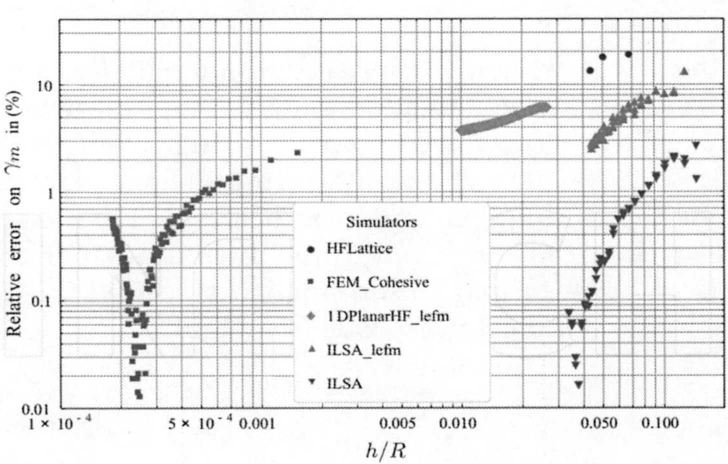

Figure 7. Evolution of the relative error in the fracture radius as a function of the mesh size for the different simulators - radial fracture in the viscosity dominated regime (i.e. zero toughness / early time, $\tau < 1$). All simulations displayed here are for $\tau < 10^{-6}$

It is interesting to investigate more closely how an algorithm using the LEFM asymptote (e.g. ILSA_lefm) is able to converge to the zero-toughness analytical solution. Consider the case of modeling a radial hydraulic fracture starting at a very small initial time (τ = 10−18), which corresponds to a dimensionless toughness K = 0.01, and is therefore very close to the M-Vertex (zero-toughness solution). If the LEFM asymptote is used in ILSA at this time, the asymptote would dictate that the fracture front needs to be advanced by roughly 104 element lengths! As already mentioned, to circumvent this problem, the ILSA_lefm code damps the front advance by rescaling the level set function T(x,y) so that the the maximum distance between any point in the ribbon elements and the damped free boundary is no more than three element lengths.

This sequence of damped front positions enables the trial widths to be relaxed until the fracture width profile presents a close approximation to the viscosity dominated solution, in spite of the fact that the tip elements are locked into the LEFM. Indeed, the algorithm settles on a solution in which the LEFM tip widths are very small and contribute very little to the net volume of the fracture, while over the remainder of the fracture away from the leading edge, the widths are locked into the viscous asymptote as dictated by the conservation of volume.

Thus the damped front advances continue until the conservation of fluid volume dictates that it should stop at which time it approximates the viscous asymptote. Results from the standard ILSA code with the correct viscous asymptote and damped ILSA_lefm code with the LEFM asymptote indicates that in order to obtain similar relative errors compared to the analytical solution, a mesh size that is an order of magnitude smaller is needed for ILSA_lefm compared to the standard ILSA code (see Figure 7).

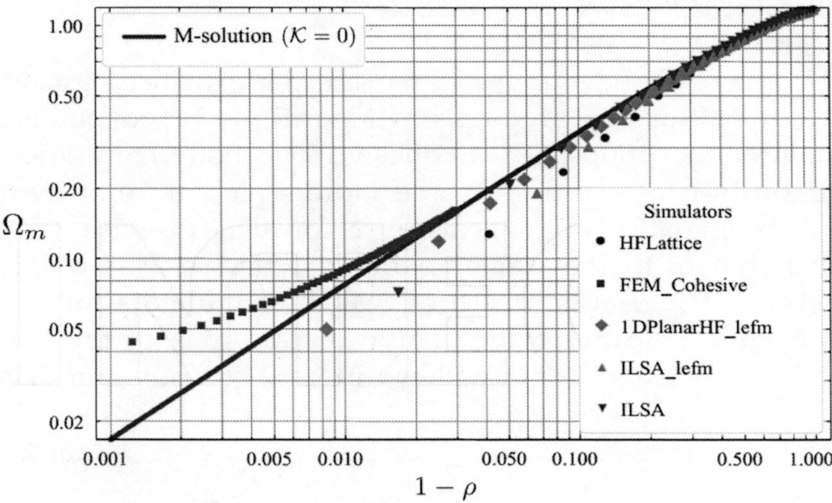

Figure 8. Fracture opening in the tip coordinates system (log-log scale) for a radial hydraulic fractures propagating in the viscosity dominated regime. Results from the different simulators for a dimensionless time $\tau < 10^{-6}$. Note that for better clarity of the plot, all the mesh points are not displayed for $1-\rho > 0.04$ for FEM_Cohesive.

Transition toward the toughness dominated regime

Finally, we investigate the performance of a subset of the algorithms on the transition of the solution toward the toughness dominated solution. Such a transition typically happens between $\tau = 1(K = 1)$ and $\tau \sim 70000(K = 3.4)$. The Explicit Moving Mesh Algorithm (EMMA), the 1DPlanarHF_lefm and the ILSA codes are compared, focusing on the evolution of the dimensionless fracture radius with time. Figure 10 display the results. The fracture radius have been averaged over the fracture footprint for the results of ILSA (3D code), while the other codes are axi-symmetric by assumption.

We clearly see that the different algorithms capture the transition between the viscosity and the toughness propagation regimes extremely well. They are virtually indistinguishable.

CONCLUSIONS

In this paper, we have investigated the performance of a number of hydraulic fracture propagation algorithms based on different propagation conditions: LEFM, cohesive zone model/tensile strength and algorithms accounting for the multi-scale nature of hydraulic fracture propagation in the near-tip region. This exercise was made possible thanks to the existence of analytical solutions for both geometries of hydraulic fractures. All the simulators investigated here are able to capture the analytical solutions of the different configurations tested, but at widely different computational costs.

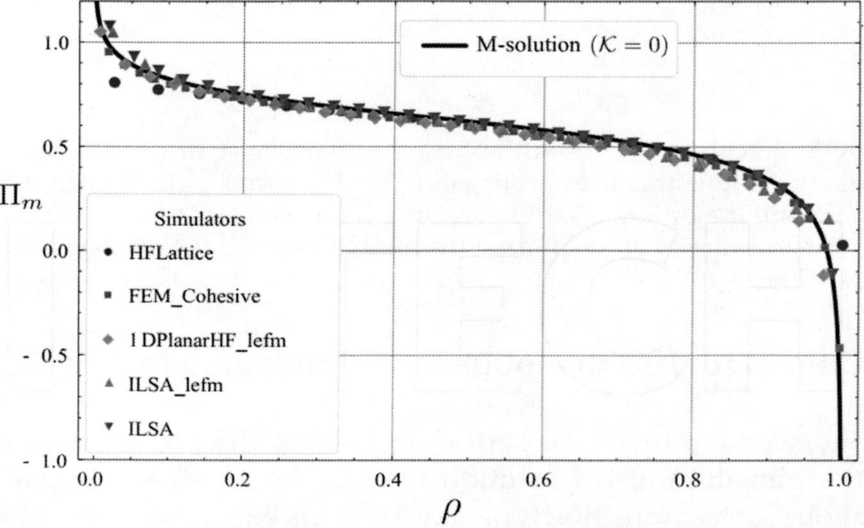

Figure 9. Dimensionless net pressure for a radial hydraulic fracture in the viscosity dominated regime (i.e. zero-toughness / early-time). Comparisons of the simulators (for a dimensionless time $\tau < 10^{-6}$) with the zero-toughness solution. The results of FEM_Cohesive and 1DPlanarHF_lefm are plotted only every 50 and 2 mesh points respectively for clarity.

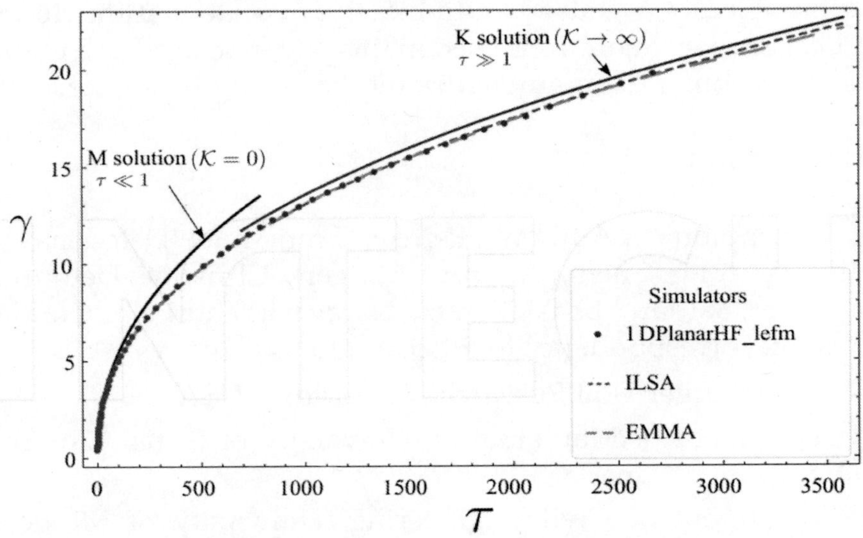

Figure 10. Radial hydraulic fracture - Transition between the viscosities dominated regime at early time to Toughness dominated regime at large time. Comparisons between different simulators and the zero (blue) and infinite toughness (red) solution.

Algorithms based on the Linear Elastic Fracture Mechanics propagation condition requires a fine mesh (h/ℓ / 10−2 – 10−3) in order to capture viscosity dominated hydraulic fracture propagation. A fine mesh is needed for these algorithms to capture the viscosity opening asymptote in order to properly match the fracture volume. Cohesive zone models, which model the fracture process zone, require even finer meshes. This is due to the fact that the cohesive zone length-scale is even smaller than that of the region of influence of the linear elastic fracture mechanics (LEFM) square-root near-tip asymptote. By contrast, when the algorithms use the appropriate multi-scale hydraulic fracture asymptote in the near-tip region, the exact solution can be matched with a very coarse mesh (i.e. h/ℓ ≈ 10−1). Extremely efficient and fast propagation algorithms can thus be developed with even better accuracy than algorithms based on the classical LEFM propagation condition. Computational cost and accuracy may not be the only concern when developing a simulator. Algorithm flexibility may also be important. We hope that the study reported in this paper can help in making an educated choice of algorithm. Finally, we also would like to advocate that the different analytical reference solutions used in this paper be used as a minimal series of benchmarks for any hydraulic fracturing simulator.

Author details

Brice Lecampion[1,*], Anthony Peirce[2], Emmanuel Detournay[3], Xi Zhang[4], Zuorong Chen[4], Andrew Bunger[4], Christine Detournay[5], John Napier, Safdar Abbas[1], Dmitry Garagash[6] and Peter Cundall[5]*
Address all correspondence to: BLecampion@slb.com

1. Schlumberger Doll Research, Cambridge, USA
2. Department of Mathematics, University of British Columbia, Vancouver, Canada
3. Department of Civil Engineering, University of Minnesota, Minneapolis, USA
4. CSIRO Earth Science and Resource Engineering, Clayton, Victoria, Australia
5. Itasca Consulting Group, Minneapolis, USA
6. Department of Civil and Resource Engineering, Dalhousie University, Canada

REFERENCES

1. R. Carbonell, J. Desroches, and E. Detournay. A comparison between a semi-analytical and a numerical solution of a two-dimensional hydraulic fracture. Int. J. Sol. Struct., 36(31):4869–4888, 1999.
2. J. Adachi and E. Detournay. Self-similar solution of a plane-strain fracture driven by a power-law fluid. International Journal for Numerical and Analytical Methods in Geomechanics, 26(6):579–604, May 2002.
3. D.I. Garagash and E. Detournay. Plane-strain propagation of a fluid-driven fracture: Small toughness solution. ASME J. Appl. Mech., 72:916–928, November 2005.
4. D. I. Garagash. Plane-strain propagation of a fluid-driven fracture during injection and shut-in: Asymptotics of large toughness. Engineering Fracture Mechanics, 73(4):456-481, March 2006.[5] J. Hu and D. I. Garagash. Plane-Strain Propagation of a Fluid-Driven Crack in a Permeable Rock with Fracture Toughness. Journal of Engineering

Mechanics, 136(9):1152, 2010.

5. A. Savitski and E. Detournay. Propagation of a penny-shaped fluid-driven fracture in an impermeable rock: asymptotic solutions. International Journal of Solids and Structures, 39(26):6311–6337, December 2002.
6. M. Madyarova. Fluid-driven penny-shaped fracture in permeable rock. Master's thesis, University of Minnesota, 2003.
7. J. Desroches, E. Detournay, B. Lenoach, P. Papanastasiou, JRA Pearson, M. Thiercelin, and A. Cheng. The crack tip region in hydraulic fracturing. Proceedings of the Royal Society of London. Series A: Mathematical and Physical Sciences, 447(1929):39, 1994.
8. D.I. Garagash and E. Detournay. The tip region of a fluid-driven fracture in an elastic medium. J. Appl. Mech., 67:183–192, 2000.
9. D. I. Garagash, E. Detournay, and J. Adachi. Multiscale tip asymptotics in hydraulic fracture with leak-off. Journal of Fluid Mechanics, 669:260–297, February 2011.
10. H. Abé, L. M. Keer, and T. Mura. "growth rate of a penny-shaped crack in hydraulic fracturing offset rocks". J. Geoph. Res., 81:6292, 1976. [12] S. Khristianovic and Y. Zheltov. Formation of vertical fractures by means of highly viscous fluids. In Proc., 4th World Petroleum Congress, volume II, pages 579–586, Rome, 1955.
11. J. Geertsma and F. De Klerk. A rapid method of predicting width and extent of hydraulically induced fractures. Journal of Petroleum Technology, 21(12):1571–1581, 1969.
12. E. Detournay. Propagation regimes of fluid-driven fractures in impermeable rocks. International Journal of Geomechanics, 4(1):35, 2004.
13. D. I. Garagash. Scaling of Physical Processes in Fluid-Driven Fracture: Perspective from the Tip. In F.M. Borodich, editor, IUTAM Symposium on Scaling in Solid Mechanics, volume 10 of Iutam Bookseries, pages 91–100, Dordrecht, 2009. Springer.
14. X. Zhang, R. G. Jeffrey, and E. Detournay. Propagation of a fluid-driven fracture parallel to the free surface of an elastic half plane. Int. J. Numer. Anal. Meth. Geomech., 29:1317–1340, 2005.
15. D. Garagash. Propagation of a plane-strain fluid-driven fracture with a fluid lag: early-time solution. Int. J. Sol. Struct., 43:5811–5835, 2006.
16. B. Lecampion and E. Detournay. An implicit algorithm for the propagation of a hydraulic fracture with a fluid lag. Comp. Meth. Appl. Mech. Engrg., 196:4863–4880, 2007.
17. Z.R. Chen. Finite element modelling of viscosity-dominated hydraulic

fractures. Journal of Petroleum Science and Engineering, 88–89:136–144, 2012.[20] E. Gordeliy and Emmanuel Detournay. A fixed grid algorithm for simulating the propagation of a shallow hydraulic fracture with a fluid lag. International Journal for Numerical and Analytical Methods in Geomechanics, 35(5):602–629, April 2011.

18. B. Lecampion. Hydraulic fracture initiation from an open-hole: Wellbore size, pressurization rate and fluid-solid coupling effects. In 46th U.S. Rock Mechanics/Geomechanics Symposium, number ARMA 2012-601, June 24–27 2012.
19. A. P. Peirce and E. Detournay. An implicit level set method for modeling hydraulically driven fractures. Computer Methods in Applied Mechanics and Engineering, 197(33-40):2858–2885, June 2008.
20. B. Damjanac, C. Detournay, P. Cundall, and Varun. Three-Dimensional Numerical Model of Hydraulic Fracturing in Fractured Rock Masses. In A. Bunger, R.G. Jeffrey, and J. McLennan, editors, Effective and Sustainable Hydraulic Fracturing, Brisbane, Australia, 20–22 May 2013. Intech.

Chapter 4

STRATEGIES TO INCREASE ENERGY EFFICIENCY OF CENTRIFUGAL PUMPS

Trinath Sahoo[1]

[1] M/S Indian Oil Corporation ltd., Mathura Refinery, India

INTRODUCTION

As the pie chart below ndicates, the LCC of a typical industrial pump over a 20 year period is primarily made up of maintenance and energy costs.

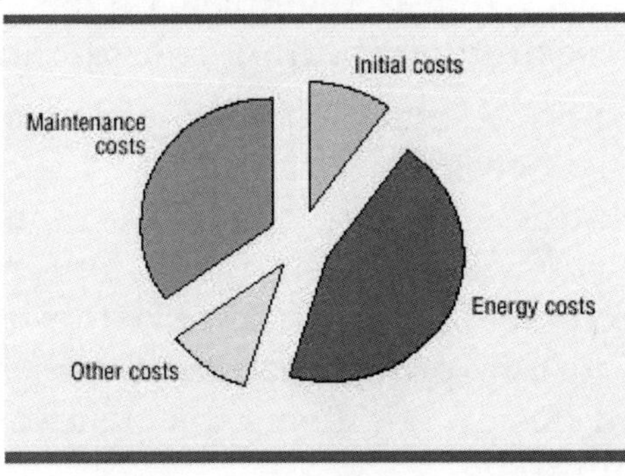

FIGURE 1. Life cycle cost of centrifugal pump.

Energy cost is highest of the total cost of owning a pump. Centrifugal pumps consume, depending on the industry, between 25 and 60% of plant electrical motor energy.

TABLE 1. Percentages of Total Motor Energy in different industries.

INDUSTRY	PUMP ENERGY (% of Total Motor Energy)
Petroleum	59%
Pulp & Paper	31%
Chemical	26%

Proper matching of pump performance and system requirements, however, can reduce pump energy costs by an average of 20 percent in many cases. The process of specifying the right pump technology for an application in facilities should go well beyond first cost, but in too many cases, it does not. Such a shortsighted approach can create major, long-term problems for organizations.

In the process industries, the purchase price of a centrifugal pump is often 5 - 10% of the total cost of ownership. Typically, considering current design practice, the life cycle cost (LCC) of a 100 horsepower pump system, including costs to install, operate, maintain and decommission, will be more than 20 times the initial purchase price. In a marketplace that is relentless on cost, optimizing pump efficiency is an increasingly important consideration.

Once the pump is installed, its efficiency is determined predominately by process conditions. The major factors affecting performance includes efficiency of the pump and system components, overall system design, efficient pump control and appropriate maintenance cycles. To achieve the efficiencies available from mechanical design, pump manufacturers must work closely with end-users and design engineers to consider all of these factors when specifying pumps.

Analysis of different losses encountered in pump operation:-

- Mechanical friction loss between fixed and rotating parts:-

  The main components are:-

  a. The external sleeve, ball or roller journal bearings.

  b. The internal sleeve bearings.

  c. Thrust bearings.

  d. The gland, neck bush and packing rings.

  The proportion of total mechanical loss contributed by each of these will depend upon the type and condition of the pump. Proper lubrication of the bearings and glands shall reduce the frictional losses.

- Disc friction loss between the liquid and the external rotating faces of the rotor discs:-

  In principle the parts concerned here are all those rotating surfaces of the pump in contact with the liquid that do not actually take share in guiding the liquid. Such elements are:-

  a. The external faces of the rotor discs or shroud.

  b. The outer edges of the shroud.

  c. The edges of the sealing rings.

  d. The whole surface of balanced discs.

  By measuring the power needed to drive the disc in a variety of conditions it appears that this power loss depends upon rotational speed, disc diameter, roughness of the sides of the discs and inner walls of the casing, the density and viscosity of the liquid and axial clearance between the disc and casing.

  The axial clearance affect the power loss. Any element of liquid in contact with revolving disk will be dragged round with it, at least for a short distance and during this journey the element will necessarily be subjected to centrifugal force. This will induce it to slide outwards. Other elements from the main body of the liquid will flow into replace the original one and hence additional kind of flow will be set up as shown in *fig. below*.

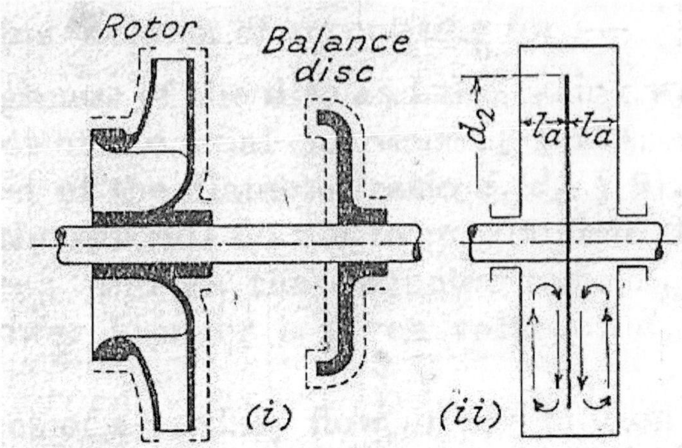

-Diagram showing areas subject to disc friction loss, etc.

FIGURE 2. Diagaram showing areas subject to disc friction loss, etc.

That is the frictional impulsion has created on a very small scale the pumping effect that the direct thrust of the impeller blades creates on an effective scale.

But it seems likely that in the space between disc and casing, the axial distance l_a will influence both the radial and tangential velocity components. If this distance is large it will be easy for relatively large amounts of liquid to become involved in the secondary circulation and thereby to steal energy from the disc. But if the distance is small, the energy should be less.

The above study shows that in comparable conditions and increase in the axial clearance causes an increase in the power loss.

Hence it can be minimized by having as small axial clearance as possible so as to improve efficiency

- The leakage power loss:-

  The leakage liquid is bled off the main stream at a number of points, each at different pressures. Leakage may occur at other region than sealing rings:-

  a. Through lantern rings of liquid sealed stuffing boxes.

  b. Past the balance disc of multistage pumps.

c. Past the neck bushes in the diaphragm of multi-stage pumps.

d. Past the main glands/mechanical seals to waste.

In normal pumps only the leakage past the balance discs of multistage pumps can be measured and the pressure head at the leak off points is fairly well known. The leakage thru the glands / mechanical seals can be reduced by attending these leakage from time to time.

- Hydraulic power loss:-

  if we have calculated the values of mechanical loss P_b, the disc friction loss P_d and leakage loss P_l, then we can assert that the residual energy $P_s - P_b - P_d - P_l$ must wholly be transferred to the main stream of liquid flowing through the pump. There is no where else for it to go. But this is a very different thing from saying that the liquid receives the corresponding net energy increment during its passage from suction flange to delivery flange. The difference between the two quantities is what we call hydraulic power loss(P_h), it can be computed in this way.

  Energy En is transferred to the liquid is utilized in imparting tangential acceleration to the liquid elements.

Unit weight of liquid will receive ($V_n V_2/g$) units of energy as there are W units of liquid effectively flowing per second it follows that

$$(W / K_p) * (V_n V_2)/g = P_s - P_b - P_d - P_l = P_w + P_h$$

where V_n is actual whirl component, V_2 is tangential velocity of liquid and k_p represents energy per second corresponding to one horse-power.Now the net energy increment per unit weight of liquid is represented by He and the net power received, W.H.P or P_w is WH_e/K_p.

Therefore hydraulic power loss Ph can be written

$$P_h=(W/K_p)*\{(V_n V_2/g)-H_e\}$$

And the hydraulic efficiency

$$\eta = [H_e/(V_n V_2/g)]=P_w/(P_w+P_h)$$

Losses under reduced and increased flow condition

An examination of inlet velocity triangle shows that additional energy losses are now to be reckoned with in increase flow or reduced flow conditions.

a. In Rotor:

If the blade tips were made tangential to the designed inlet relative velocity vr1, they cannot be at the same time be tangential to the modified velocities vr1c or vr1d. Because of this eddies may form.

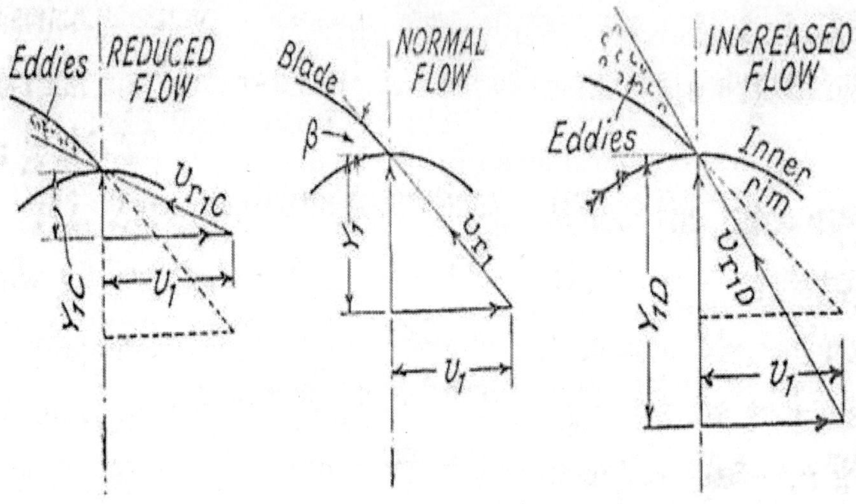

FIGURE 3. Normal and distorted inlet velocity diagrams.

When the state of flow is below normal, there may be a tendency for the liquid stream in each of impeller passages to concentrate near the front of the blades, leaving a more or less dead space near the back of the blades, that in regard to effective forward motion but alike and its capacity to waste energy.

- In the Recuperator :-

When the pump discharge is reduced there will be additional energy losses. While the velocity of the liquid leaving the impeller and entering the volute is above normal yet the mean velocity in the volute itself must be below normal. Thus the essential condition for maximum efficiency of energy conversion can no longer apply.

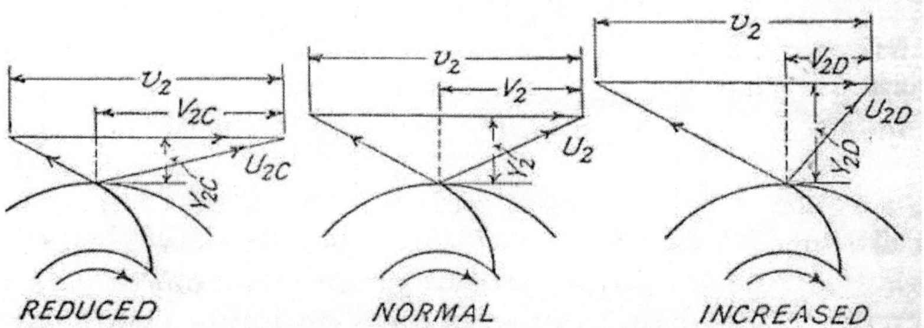

FIGURE 4. Ideal outlet velocity diagrams for reduced, normal, and increase flow in centrifugal pump impeller.

If now the pump discharge is considerably above normal, then the absolute velocity vector (figure 1) takes an abnormal inclination. Very serious contraction of main stream of liquid may occur at volute tongue (figure 2) with consequent energy losses.

In regard to guide blade or diffuser type recuperator(figure 5) shows that abnormally higher rates of flow will again destroy the necessary correspondence between vane inclination and velocity vector inclination. Energy dissipation is augmented here also

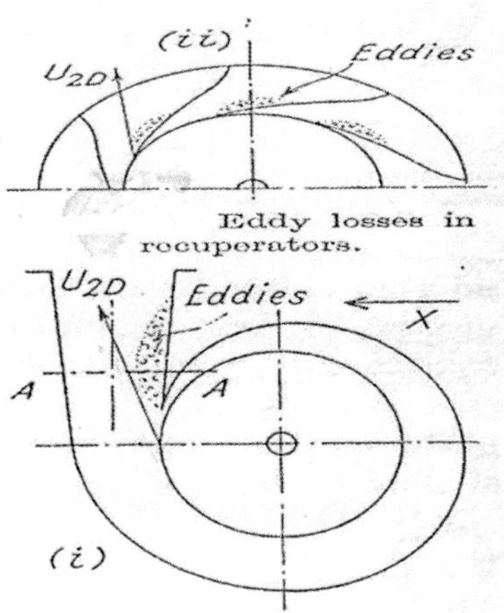

FIGURE 5. As mentioned in text.

OPERATION AT HIGH FLOWS

The vast majority of pumping systems run far from their best efficiency point (BEP). For reasons ranging from shortsighted or overly conservative design, specification and procurement to decades of incremental changes in operating conditions, most pumps, pipes and control valves are too large or too small. In anticipation of future load growth, the end-user, supplier and design engineers routinely add 10 to 50% "safety margins" to ensure the pump and motor can accommodate anticipated capacity increases.

Under this circumstances the head capacity curve intersects the system head curve at a capacity much in excess of the required flow using excess power. Of course the pump can be throttled back to the required capacity and the power is reduced some what. But if the pump runs uncontrolled it will always run at excess flow. unless sufficient NPSH has been made available, the pump may suffer cavitation damage and power consumption will be excessive. Important energy savings can be made if at the time of selecting the condition of service, reasonable restraints are exercised to avoid using excessive safety margins for obtaining the rated service condition. But in an existing installation if the pumps have excessive margins the following options are available.

1. The existing impeller can be cut down to meet the condition of service required for the installation.
2. A replacement impeller with the necessary reduced diameter may be ordered from the pump manufacturer.
3. In certain cases there may be two separate impellers designs available for the same pump, one of which is of narrower width than the one originally furnished. A narrower replacement will have its best efficiency at at a lower capacity than the normal width impeller.

EFFECT OF SPEED AND ITS CONTROL

Speed of the centrifugal pump has marked effect on the varios losses taking place in the pump and hence it is also one the important factor which can be controlled tyo improve efficiency.

EFFECT OF SPEED ON VARIOUS LOSSES

Mechanical losses: In the pump the correlation between speed N and mechanical power loss P_b will be

$$P_b = K_1 . N^m$$

The value of m is between 1 & 2depending upon the type of bearing and stuffing box.

The input power varies as the cube of speed.

$$P_s = K_2 . N^3$$

So the relative mechanical loss $= P_b/P_s = K_1 . N^m / K_2 . N^3$

Disc friction loss: For a given pump using liquid the relation between speed and power loss can be given by

$$P_d = K_3 . N^{2.85}$$

The relative disc friction loss $= P_d / P_s = K_3 . N^{2.85} // K_2 . N^3$

It show that the value relative disc friction loss increases as the speed decreases.

Leakage: The relative leakage loss appears to be independent of the pump speed.The leakage takes place. It seems likely that the relative leakage power loss will slightly increase as the pump speed falls.

Hydraulic Power loss: As the liquid flows at a rate Q through the pump passage, it undergoes a energyu loss consist of friction loss and eddy losses.

The hydraulic power loss is denoted by

$$P_h = K_{11} . N^{2.95}$$

Comparing this with the input power

$$\Delta n = K_{11} . N^{2.95} / K_2 N^3$$

Hence the value of the relative power loss rises as the pump speed falls.

These kind of losses can be minimized by having a proper speed control mechanism.

Speed control is an option that can keep pumps operating efficiently over a broad range of flows. In centrifugal pumps, speed is linearly related to flow but has a cube relationship with power. For example, slowing a pump from 1800 to 1200 rpm results in a 33%

decreased flow and a 70% decrease in power. This also places less stress on the system.

There are two types of speed control in pumps: multi-speed motors and variable speed drives. Multi-speed motors have discrete speeds (e.g., high, medium, and low). Variable speed drives provide speed control over a continuous range. The most common type is the variable frequency drive (VFD), which adjusts the frequency of the electric power supplied to the motor. VFDs are widely used due to their ability to adjust pump speed automatically to meet system requirements. For systems in which the static head represents a large portion of the total head, however, a VFD may be unable to meet system needs.

VARIABLE SPEED DRIVES

Pump over-sizing causes the pump to operate to the far left of its best efficiency point (BEP) on the pump head -capacity curve. Variable speed drives, assuming a low static head system, allow the pump to operate near its best efficiency point (BEP) at any head or flow. In addition, the drive can be programmed to protect the pump from mechanical damage when away from BEP -- thereby enhancing mechanical reliability. Furthermore, excessive valve throttling is expensive and not only contributes to higher energy and maintenance cost, but can also significantly impair control loop performance. Employing a throttled control valve, less than 50% open, on the pump discharge may accelerate component wear, thereby slowing valve response.

VFDs allow pumps to run at slower speeds with further contributions to pump reliability and significant improvement in mean-time-between-failure (MTBF).

EFFECT OF SPECIFIC SPEED

The higher the specific speed selected for a given set of operating condition, the higher the pump efficiency and therefore the lower the power consumption. Barring other considerations the tendency should be to favour higher specific speed selection from the point of view of energy conservation

CLEARANCE

Good wear ring with proper clearance improves pump reliability and reduce energy consumption. Correct impeller to volute or back plate clearance is also to be maintained. Pump efficiency decreases with time because of wear. A well designed pump usually comes with a diametral clearance of 0.2 to 0.4%. however as long as it remains below 0.6 to 0.8 % its effect on efficiency remains negligible. When the clearance start to increase beyond these values efficiency start to drop drastically. For equal operating condition the rate of wear depends primarily on design and material of the wear ring. Generally for non corrosive liquids, the resistance to wear increases with hardness of the sealing surface material.

If a pump having a specific speed of 2500, the leakage loss in a new pump will be about 1%. Thus when the internal clearances will have increased to the point that this leakage will have doubled, we can regain approximately 1% in power saving by restoring the pump clearance. But if we are dealing with a pump having a specific speed of 750, it will have leakage loss of about 5%. If the clearances are restored after the pump has worn to the point that its leakage losses have doubled, we can count on 5% power saving.

CHANGE IN SURFACE ROUGHNESS

Depending on the material of construction and properties of the liquid being pumped, the roughness of the flow path can also change over time. In some instances, the channels may acquire a smooth polish, and in other they may become roughened. Both of these changes can significantly affect pump performance. An increase in casing roughness usually reduces both the total head and the efficiency.

CHANGE IN FLOW PATH SIZE

The dimension of the pump's flow path may change over time due to abrasion or erosion, which usually increases the pathways' dimension or to scale, rust or sedimentation which usually reduces the size of the pathways. The latter is particularly apt to occur in pumps operating intermittently.

RUN ONE PUMP INSTEAD OF TWO

Many installations are provided with two pumps operating in parallel to deliver the required flow under full load. Too often, both pumps are kept on line even when demand drops to a point where a single pump can carry the load. The amount of energy wasted in running two pumps at half load when a single pump can meet this condition is significant.

If we want to reduce the flow to half load and still maintain both pumps online, it will be necessary to throttle the pump discharge and create a new system head curve. Under these condition, each pump will deliver 50% of the rated capacity at 117% rated head, much of which will have to be throttled. Each pump will take 72.5% of its rated power consumption. Thus the total power consumption of two pumps of two pumps operating under half load condition would be 145% of that required if a single pump were to be kept on line.

VISCOSITY OF LIQUID

Liquid viscosity affect pump performance. This is because two of the major losses in a centrifugal pump are caused by fluid friction and disc friction. These losses vary with the viscosity of the liquid being pumped, so that both the head capacity output and the mechanical output differ from the original values. As viscosity of the liquid increases the head developed by the pump decreases and the efficiency decreases. So in process industries it is required to maintain good insulation and steam tracing in the suction line of the pump.

EFFECT OF CAVITATION

Due to cavitation the vapor bubbles are formed and these bubbles will modify the velocity distribution as well as the pressure distribution in the rotor passage. The effect of vapor pressure bubbles will virtually lower the density of the liquid; that is the pump behaves as if another liquid of less density were flowing through it. As the mean velocity of the fluid mixture is now at higher (for the weight per second is unchanged) the outlet velocity diagram tends to assume the distorted shape and shall lower the total head. The fall in density reduces the pressure generated.

Due to cavitation there will be loss of energy.

EFFECT OF CHANGE OF DENSITY

The pump develops the same head in meters of liquid independent of specific gravity. The pump delivers the same quantity by volume independent of specific gravity but the quantity by weight will be proportional to specific gravity. The input power and the output power for a given volumetric discharge vary in proportion to density.

If the water is heated the density changes. In high pressure steam boilers, the reduction in density of the water may be 15 percent or more. If a given weight per second of water is to be discharged against a stipulated pressure, the power output will increase as the water temperature increases.

Pump Sizing

Selecting a centrifugal pump can be challenging because these pumps generate different amounts of flow at different pressures. Each centrifugal pump has a "best efficiency point" (BEP). Ideally, under normal operating conditions, the required flow rate will coincide with the pump's BEP.

The complexity associated with selecting a pump often results in a pump that is improperly sized for its application. Selecting a pump that is either too large or too small can reduce system performance. Undersizing a pump may result in inadequate flow, failing to meet system requirements. An oversized pump, while providing sufficient flow, can produce other negative consequences; higher purchase costs for the pump and motor assembly; higher energy costs, because oversized pumps operate less efficiently; and higher maintenance requirements, because as pumps operate further from their BEP they experience greater stress; ironically, many oversized pumps are purchased with the intent of increasing system reliability.

Unfortunately, conservative practices often prioritize initial performance over system life cycle costs. As a result, larger-than-necessary pumps are specified, resulting in systems that do not operate optimally. Increased awareness of the costs of specifying oversized pumps should discourage this tendency.

Variable Loads

In systems with highly variable loads, pumps that are sized to handle the largest loads may be oversized for normal operating loads. In these cases, the use of multiple pumps, multi-speed motors, or variable speed drives often improves system performance over the range of operating conditions.

To handle wide variations in flow, multiple pumps are often used in a parallel configuration. This arrangement allows pumps to be energized and de-energized to meet system needs. One way to arrange pumps in parallel is to use two or more pumps of the same type. Alternatively, pumps with different flow rates can be installed in parallel and configured such that the small pump—often referred to as the "pony pump"—operates during normal conditions while the larger pump operates during periods of high demand.

VALVES AND FITTINGS

Pumping system controls should be evaluated to determine the most economical control method. High-head-loss valves, such as globe valves, are commonly used for control purposes. Significant losses occur with these types of valves, however, even when they are fully open. If the evaluation shows that a control valve is needed, choose the type that minimizes pressure drop across the valve. Pumping system control valve inefficiencies in plant processes offer opportunities for energy savings and reduced maintenance costs. Valves that consume a large fraction of the total pressure drop for the system, or are excessively throttled, can be opportunities for energy savings. Pressure drops or head losses in liquid pumping systems increase the energy requirements of these systems. Pressure drops are caused by resistance or friction in piping and in bends, elbows, or joints, as well as by throttling across the control valves. The power required to overcome a pressure drop is proportional to both the fluid flow rate (given in gallons per minute [gpm]) and the magnitude of the pressure drop (expressed in ft of head). The friction loss and pressure drop caused by fluids flowing through valves and fittings depend on the size and type of pipe and fittings used, the roughness of interior surfaces, and the fluid flow rate and viscosity.

Case study 1 –

In the Hydrocracker Unit of a petroleum refinery the reflux pump generally pumps Naptha with a specific gravity of 0.7. The motor KW rating was 190 kw. Due to some adverse situation in the process, sometimes water also comes in the stream resulting in increase of specific gravity of 0.9.As power P=(wQH/3960) where w is specific weight, Q is discharge and H is the manometric head As a result the motor draws more current. If it goes unnoticed the energy will be wasted.

Case study 2 -

The pump develops the same head in meters of liquid independent of specific gravity. The pump delivers the same quantity by volume independent of specific gravity but the quantity by weight will be proportional to specific gravity. The input power and the output power for a given volumetric discharge vary in proportion to density.

In the Vacuum Distillation Unit of a petroleum refinery the column bottom pump pumps Short residue with a specific gravity of approximately 0.9. Most of the refinery are configured for different type of crudes, like high sulphur and low sulphur. During processing of high sulphur crude the specific gravity and viscosity of the short residue increases. This results in more power consumption.

Case study 3 -

In systems with highly variable loads, pumps that are sized to handle the largest loads may be oversized for normal operating loads. In these cases, the use of multiple pumps, multi-speed motors, or variable speed drives often improves system performance over the range of operating conditions.

To handle wide variations in flow, multiple pumps are often used in a parallel configuration. This arrangement allows pumps to be energized and de-energized to meet system needs. One way to arrange pumps in parallel is to use two or more pumps of the same type. Alternatively, pumps with different flow rates can be installed in parallel and configured such that the small pump operates during normal conditions while the larger pump operates during periods of high demand.

In the Crude unit of a petroleum refinery the main crude pumps are installed in parallel. Some times the crude unit thru put depends upon the secondary unit thruput. If there is any disturbance in the secondary units, then the load in the Crude unit has to brought down. In this case multiple pumps are used in a parallel configuration. This arrangement allows saving energy.

Case study 4 -

The dimension of the pump's flow path may change over time due to scale, rust or sedimentation which usually reduces the size of the pathways. The latter is particularly apt to occur in pumps operating intermittently.

The sulphur pit pump in a Sulphur recovery unit operates intermittently. It pumps liquid sulphur in a jacketed pipe. If the steam in the jacket is not circulated properly or due to ageing, some sulphur particles get deposited in the line. This reduces the flow path size. An increase in pipe roughness increases the friction loss and usually reduces the efficiency. Because of these reasons the pump started drawing more power and tripped more frequently. So to avoid the deposition of sulphur in the line steam was injected so as to clean the sulphur line.

CONCLUSION

The following are different ways to conserve the Energy in Pumping System:

- When actual operating conditions are widely different (head or flow variation by more than 25 to 30%) than design conditions, replacements by appropriately sized pumps must be considered.
- Operating multiple pumps in either series or parallel as per requirement.

- Reduction in number of pumps (when System Pressure requirement, Head and Flow requirement is less).
- By improving the piping design to reduce Frictional Head Loss
- By reducing number of bends and valves in the piping system.
- By avoiding throttling process to reduce the flow requirement.
- By Trimming or replacing the Impellers when capacity requirement is low.
- By using Variable Speed Drives

REFERENCES

1. W. Driedger, 1995 "Controlling Centrifugal Pumps", Hydrocarbon Processing, July 1995
2. T. F. Glover, 1975 "Understanding NPSH for Pumps", Technical Publishing Co. 1975,
3. I. J. Karassik, 1987 "Centrifugal pumps operation at off-design conditions", Chemical Processing April,May, June 1987,
4. N. P. Lieberman, 1991 "Trouble shooting Process Operations", 3rd Edition, PennWell Books
5. Refining Department 1981 "Centrifugal Pumps for General Refinery Services", API Standard 610, 6th Edition, January 1981

Chapter 5

FRACTURES AND FRACTURING: HYDRAULIC FRACTURING IN JOINTED ROCK

Charles Fairhurst[1, 2]

[1] Senior Consultant, Itasca Consulting Group, Inc, Minneapolis Minnesota, USA

[2] Professor Emeritus, University of Minnesota, Minneapolis, Minnesota, USA

Rock in situ is arguably the most complex material encountered in any engineering discipline. Deformed and fractured over many millions of years and different tectonic stress regimes, it contains fractures on a wide variety of length scales from microscopic to tectonic plate boundaries.

Hydraulic fractures, sometimes on the scale of hundreds of meters, may encounter such discontinuities on several scales. Developed initially as a technology to enhance recovery from petroleum reservoirs, hydraulic fracturing is now applied in a variety of subsurface engineering applications. Often carried out at depths of kilometers, the fracturing process cannot be observed directly.

Early analyses of the hydraulic fracturing process assumed that a single fracture developed symmetrically from the packed off-pressurized interval of a borehole in a stressed elastic continuum. It is now recognized that this is often not the case. Pre-existing fractures can and do have a significant influence on fracture development, and on the associated distributions of increased fluid pressure and stresses in the rock. Given the usual lack of information and/or uncertainties concerning important variables such as the disposition and mechanical properties of pre-existing fracture systems and properties, rock mass permeabilities, in-situ stress state at the depths of interest, fundamental questions as to how a propagating fracture is affected by encounters with pre-existing faults, etc., it is clear that design of hydraulic fracturing treatments is not an exact science.

Fractures in fabricated materials tend to occur on a length of scale that is small; of the order of the 'grain size' of the material. Increase in the size of the structure does not introduce new fracture sets.

Numerical modeling of fracture systems has made significant advances and is being applied to attempt to assess the extent of these uncertainties and how they may affect the outcome of practical fracturing programs. Geophysical observations including both micro-seismic activity and P- and S-wave velocity changes during and after stimulation are valuable tools to assist in verifying model predictions and development of a better overall understanding of the process of hydraulic fracturing on the field scale. Fundamental studies supported by laboratory investigations can also contribute significantly to improved understanding.

Given the widening application of hydraulic fracturing to situations where there is little prior experience (e.g., Enhanced Geothermal Systems (EGS), gas extraction from 'tight shales' by fracturing in essentially horizontal wellbores, etc.) development of a greater understanding of the mechanics of hydraulic fracturing in naturally fractured rock masses should be an industry-wide imperative. HF 2013 International Conference for Effective and Sustainable Hydraulic Fracturing is very timely!

This lecture will describe examples of some current attempts to address these uncertainties and gaps in understanding. And, it is hoped, it will stimulate discussion of how to achieve more effective practical design of hydraulic fracturing treatments.

INTRODUCTION

The term 'rock' covers a wide variety of materials and widely different rheological properties often proximate to each other in the subsurface. Tectonic and gravitational forces, sustained over millions of years, have deformed and fractured the rock on many scales. These forces are transmitted in part through the solid skeleton of the rock, and in part through the fluids under pressure in the pore spaces. Long-term circulation through rock at high temperatures at depth involves dissolution and precipitation along the fluid pathways, producing changes in the chemical composition of the fluids and modifying the overall fluid circulation.

Rock in situ is 'pre-loaded' and in a state of changing equilibrium. Any engineering activity changes this equilibrium (see Appendix 1). Often the changes can be accommodated in stable fashion, but serious instabilities can develop.

The rock mass is opaque. Although geophysics is making impressive advances in defining large structures such as faults and bedding planes, most of the features that influence the rock response to engineering activities remain hidden. Mining and civil engineering activities allow three-dimensional access to the underground and direct observation of smaller features such as fracture networks, but most of the newer engineering applications involve essentially one-dimensional access by borehole. Rock engineering problems fall into the 'data -limited' category, as defined by Star field and Cundall (1988), and strategies to address them must follow a different strategy than engineering problems where detailed and precise design information is available.

Faced with such complexity and lack of structural details, traditional subsurface engineering design has been guided by empirical procedures developed and refined through long experience.

Projects are now venturing well beyond current experience, and for many, 'novel' applications now considered (e.g., Enhanced Geothermal Systems, Carbon Sequestration, see Appendix 1). There is little experience, few guiding rules and very little data to guide the engineering approach.

Such obstacles notwithstanding, subsurface processes, both long-term geological and short term responses, to engineering activities do obey the laws of Newtonian Mechanics. Classical continuum mechanics has long been used to guide some aspects of design, but considerable care is required in practical application, due to the need to simplify the representation of the real conditions in order to obtain analytical solutions.

The remarkable developments in high-speed computation and associated modeling techniques over the past one to two decades provide an important new tool, which complemented by the appropriate field instrumentation, can augment the classical continuum analyses and help overcome the lack of prior experience. Some empiricism and general practical guidelines may still be useful for the design engineer, but these can and should be mechanics-informed.

This lecture attempts to illustrate the 'mechanics-informed' approach with respect to the practical application of hydraulic fracturing and related engineering procedures to rock engineering.

HYDRAULIC FRACTURING

Hydraulic fracturing first was used successfully in the late 1940's to increase production from petroleum reservoirs (Howard and Fast, 1970). The technology has evolved since and is now a major, essential technique in oil and gas production. This and other impressive oil industry developments, such as directional drilling, have attracted interest in application of these technologies to a variety of other subsurface engineering operations. Enhanced Geothermal Energy (EGS) is a notable example. Geothermal Energy is a huge resource. Commenting on the EGS resource in the USA, Tester et al. (2005), state:

"....we have estimated the total EGS resource base to be more than 13 million exajoules (EJ)

[1] Using reasonable assumptions regarding how heat would be mined from stimulated EGS reservoirs, we also estimated the extractable portion to exceed 200,000 EJ or about 2,000 times the annual consumption of primary energy in the United States in 2005. With technology improvements, the economically extractable amount of useful energy could increase by a factor of 10 or more, thus making EGS sustainable for centuries."

[2] -"At this point, the main constraint is creating sufficient connectivity within the injection and production well system

in the stimulated region of the EGS reservoir to allow for high per-well production rates without reducing reservoir life by rapid cooling."

[3] -Field experiments to extract geothermal energy from rock at depth by hydraulic fracturing were started in 1970 by scientists of the Los Alamos National Laboratory, USA. Two boreholes were drilled into crystalline rock (one 2.8 km deep, rock temperature 195°C; the other 3.5 km rock, 235°C) at Fenton Hill, New Mexico. Hydraulic fracturing was used to develop fractures from the boreholes in order to create a fractured region through which water could be circulated to extract heat from the rock. The experiment was terminated in 1992. Commenting on what was learned from the Fenton Hill study, Duchane and Brown (2002) note:

"The idea that hydraulic pressure causes competent rock to rupture and create a disc-shaped fracture was refuted by the seismic evidence. Instead, it came to be understood that hydraulic stimulation leads to the opening of existing natural joints that have been sealed by secondary mineralization. Over the years additional evidence has been generated to show that the joints oriented roughly orthogonal to the direction of the least principal stress open first, but that as the hydraulic pressure is increased, additional joints open."

This is an early indication that pre-existing fractures mass significantly affect how hydraulic fractures propagate in a rock mass.

INFLUENCE OF FRACTURES AND DISCONTINUITIES ON THE STRENGTH OF BRITTLE MATERIALS

Hydraulic fracturing can be considered as a technique to overcome the strength of a rock mass in situ, initiation and propagation of a crack through a system of pre-existing fractures, essentially planar discontinuities (e.g., bedding planes), and intact rock.

In examining the fracture propagation process, the pioneering work of Griffith (1921, 1924) is a logical point of departure. Griffith had identified planar discontinuities, or flaws, in fabricated materials as the reason why the observed technical strength of brittle materials was about three orders of magnitude lower than the theoretical inter-atomic cohesive (tensile) strength.[4] - Using an analytical solution by Inglis (1913) for the elastic stresses generated around an elliptical crack in a plate, Griffith observed that the maximum tensile stress at the tip of the crack $\sigma_t = \sigma_0$ (1+ 2a/b), where a and b are the major and minor semi-axes of the ellipse, and as the ellipse degenerated to a sharp crack or flaw (i.e., as the ratio a/b became very high)[5] - , the stress σ_t could rise to a value high enough to reach the inter-atomic cohesive strength sufficient to cause the original crack to start to extend. But would the crack continue to extend and lead to macroscopic failure? To address this question, Griffith invoked the Theorem of Minimum Potential Energy, which may be stated as "The stable equilibrium state of a system is that for which the potential energy of the system is a minimum." For the particular application of this theorem to brittle rupture, Griffith added the statement, "The equilibrium position, if equilibrium is possible, must be one in which rupture of the solid has occurred, if the system can pass from the unbroken to the broken condition by a process involving a continuous decrease of potential energy."[6] -

Griffith's classical work has provided the foundation for the field of "Fracture Mechanics" [Knott (1973); Anderson (2005)] responsible for major continuing advances in the development of high-performance fabricated materials.

Since we will make reference later to this specific definition by Griffith, it is useful to re-state it here.

THEOREM OF MINIMUM POTENTIAL ENERGY

"The stable equilibrium state of a system is that for which the potential energy of the system is a minimum. The equilibrium position, if equilibrium is possible, must be one in which rupture of the solid has occurred, if the system can pass from the unbroken to the broken condition by a process involving a continuous decrease of potential energy."

Although much of classical Fracture Mechanics has emphasized applications to problems of Linearly Elastic Fracture Mechanics (LEFM) it is important to recognize that the theorem of minimum potential applies equally to inelastic problems.

MECHANICS OF HYDRAULIC FRACTURING

As used classically in petroleum engineering, hydraulic fracturing involves sealing off an interval of a borehole at depth in an oil or gas bearing horizon, subjecting the interval to increasing fluid pressure until a fracture is generated, injecting some form of granular prop pant into the fracture as it extends a considerable distance from the borehole into the petroleum bearing formation, and then releasing the pressure. This causes the sides of the fracture to compress onto the prop pant, creating a high-permeability pathway to allow oil and/or natural gas to flow back to the well and to the surface.

Figure 1 shows a simple two-dimensional cross-section through an idealized hydraulic fracture. The borehole injection point is at the center of the fracture, which is assumed to be a narrow ellipse that has extended in a plane normal to the direction of the maximum [7] - (least compressive) in-situ stress.

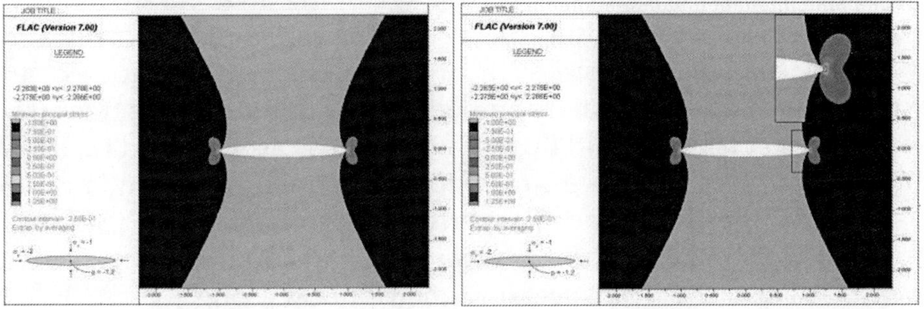

FIGURE 1. (Left) Major and (right) minor principal stresses in the vicinity of an internally pressurized elliptical crack in an impermeable rock.

In the case shown, the crack major/minor axis ratio a/b is 10:1. The internal fluid pressure p = 1.2, while the least compressive principal stress $\sigma x = 1.0$. This results in a tensile stress concentration at the crack tip. The magnitude of the elastic stress concentration at the crack tip increases directly with 2a/b, (Inglis, 1913). Hence for the case of a>>b, i.e., a 'sharp' crack[8] - , the concentration is very high, and the crack will extend essentially as soon as the fluid pressure exceeds the magnitude of the least compressive principal stress (σx in Figure 3) it begins to extend, and there will be a pressure gradient from the injection point towards the crack tip as the fluid flows towards the tips. This gradient will depend on the fluid viscosity. Also, since the rock will exhibit some level of permeability, fluid will also flow (or 'leak–off') into the formation as it flows under pressure along the fracture; the rock has a finite strength, or 'toughness' so that energy will be required to extend the crack.

An analytical solution for the stresses in the elastic medium and the crack-opening displacement along the crack was first published by Inglis (1913) and served as the basis for early applications to hydraulic fracturing and fracture treatment design. The Perkins, Kern (1961) and Nordgren (1972) (PKN) and Geertsma and de Klerk (1969) (GDK) models are still used, although numerical models and combinations are now popular. Details of the PKN and GDK models can be found on the SPE website: http://petrowiki.spe.org/Fracture_propagation_models. Several differences between the stationary crack assumed by Inglis (1913) and a hydraulic fracture introduce significant difficulties in developing an accurate model of the fracturing process. Thus, the fracture is generated by application of an increasing fluid pressure until the fracture is initiated and extends away from the injection point. Flow of fluid in the fracture is governed by classical fluid flow equations of Poiseuille and Reynolds (lubrication); the pressure drop along the fracture depends on the viscosity of the fluid, and the permeability of the rock (leading to fluid 'leak-off'); the fracture aperture depends on the stiffness of the rock mass and the fluid pressure distribution along the crack; and fracture extension depends on the mechanical energy supplied to the region around the crack tip. The tip may propagate ahead of the fluid, leading to a 'lag,'a dry region between the crack tip and fluid front.

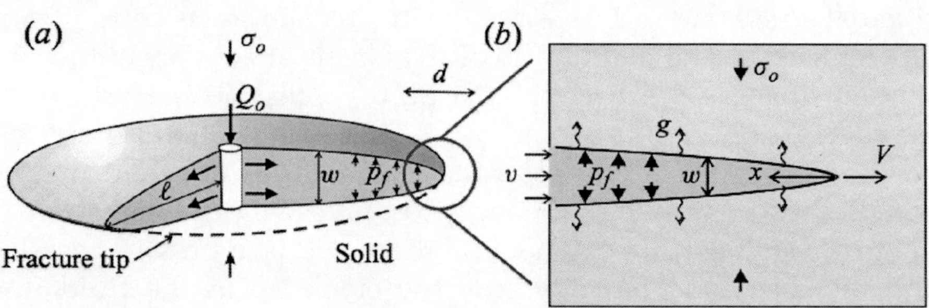

FIGURE 2. Radial Model of Axi-symmetric Flow and Deformation associated with Hydraulic Fracturing.

Figure 2 illustrates these features for the classical Radial Model in which it is assumed that the fracture propagates symmetrically away from the borehole in a plane normal to the minimum (least compressive) principal in-situ stress, σ0.

Development of efficient and robust Hydraulic Fracturing (HF) simulators is central to successful practical HF treatment of petroleum reservoirs. As noted earlier, competing physical processes are operative during the fracturing operation. This has led to a sustained effort over many years to understand and map the multi-scale nature of the tip asymptotics that arise as a result of these competing physical processes in fluid-driven fracture. These asymptotics solutions are critical to the construction of efficient and robust HF simulators. For example, in an impermeable medium, the viscous energy dissipation associated with driving fluid through the fracture competes with the energy required to break the solid material. Breaking of the bonds corresponds to the familiar asymptotic form of linear elastic fracture mechanics (LEFM), i.e., the opening in the tip region is of the form, e.g., (Rice, 1968), with denoting the distance from the tip. However, under conditions where viscous dissipation dominates, the coupling between the fluid flow and solid deformation leads to (Spence and Sharp, 1985; Lister, 1990; Desroches et al., 1994), on a scale that is considerably larger than the size of the LEFM-dominated region, but still small relative to the overall fracture size. In other words, in the viscosity-dominated regime, the zone governed by the LEFM asymptote is negligibly small compared to the crack length. Thus, in the viscosity-dominated regime, the HF simulator should embed a 2/3 power law asymptote rather than the classic 1/2 asymptote of LEFM. Garagash et al.(2011) discuss the generalized asymptotic near the tip an advancing hydraulic fracture, an extension of two particular asymptotic obtained at Schlumberger Cambridge Research Laboratory in the early 1990′s (Desroches et al., 1994; Lenoach, 1995).

Three classes of numerical algorithms for HF simulators have now been built: (i) a moving grid for KGD, radial, PKN and P3D fracture simulators; (ii) a fixed grid for plane strain and axisymmetric HF with allowance for a lag between the fluid front and the crack tip, and fracture curving (a versatile code has been developed at CSIRO[9]

- Melbourne to simulate the interaction of a hydraulic fracture with other discontinuities); and (iii) fixed grid for simulating a arbitrary shape planar fracture in a homogenous elastic rock. These codes rely on the displacement discontinuity method (Crouch and Star field, 1983) for solving the elastic component of the problem, i.e., the relationship between the fracture aperture and the fluid pressure.

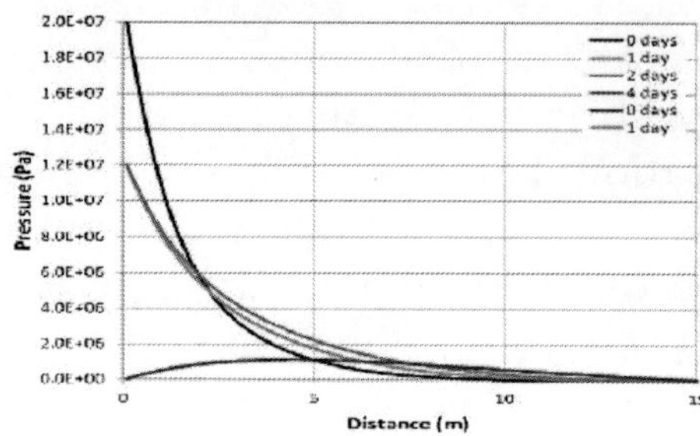

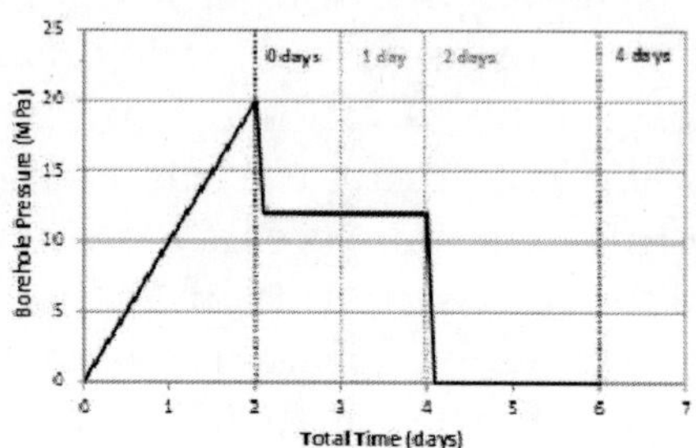

FIGURE 3.Fluid Pressure Distribution along the Central Axis (Ox) of Figure 1 for a permeable rock due to pressurization and de-pressurization of the bore-hole.

Figure 3 is presented to illustrate that the fluid pressure in a permeable rock can continue to flow away from the point of injection even after the borehole pressure is reduced to zero. The example shows the distribution of fluid pressure in the rock mass (permeability 5 mD) after (i) 2 days of pressurization up to the peak pressure of 20 MPa in the fracture; (ii) stop pumping and reduce fluid pressure quickly to 12MPa at the point of injection; (iii) hold the pressure constant for 2 days; and (iv) drop the pressure to zero.

It is seen that the pressure in the rock (red curve) has a maximum at some distance from the borehole such that fluid continues to flow into the rock for some time after the pressure in the borehole is reduced to zero. Different combinations of rock permeability, pumping rates and durations can lead to higher peak pressure values in the rock, and longer periods during which fluid can continue to flow away from the well. Such flow may contribute to slip on pre-existing fractures after the pressure in the borehole is reduced to zero.

HYDROSHEAR

Hydraulic fracturing is considered to be initiated from a packed–off interval borehole when the net state of stress around the well bore reaches the tensile strength of the rock. It is important to recognize that fluid pressurization of a well in permeable rock will result in flow of the fluid into the rock as soon as the fluid pressure stimulation process is started. This changes the effective stress state in the rock mass and can lead to slip on pre-existing fractures at fluid pressures below the pressure required to crate and extend a hydraulic fracture. This process of inducing slip on pre-existing fractures is termed 'Hydro-shear'. Flow of pressurized fluid into the rock reduces the effective normal stress (σn - p) everywhere in the rock { σn = normal stress at any point; p = fluid pressure.] If c and μ respectively represent the cohesion and coefficient of friction acting across the surfaces of a fracture in the rock, then the effective resistance of the fracture to (shear) sliding, τr, will be:

$$\tau r = c + \mu\ (\sigma n - p)$$

Thus, if the pressure p is raised progressively then τr will be reduced correspondingly until it reaches the limit at which sliding will occur. The situation is illustrated graphically in Figure 3. The rock is subjected to a three-dimensional state of stress represented by the principal stresses $\sigma 1$, $\sigma 2$, $\sigma 3$ and the fluid pressure p. The series of points 'X' indicate the effective state of stress on an array of pre-existing fractures in the rock. As illustrated in Figure 5, the effect of increasing the fluid pressure in the medium is to move the stress state on these cracks close to the limiting shear resistance, i.e., to the limiting value represented by the Mohr-Coulomb limit. As the stress state reaches this limit, the cracks will slip. In order to initiate a hydraulic fracture, the fluid pressure would need to be increased further, until the limiting Mohr circle reaches the tensile strength limit of the failure envelope. Since crack surfaces are often not smooth, shear slip will tend to result in crack dilation, and an associated increase in fluid conductivity. It is suggested that hydro-shearing could be more effective than hydraulic fracturing as a stimulation technique in certain applications, e.g., in stimulation of high-temperature geothermal reservoirs. Cladouhos et al. (2011) discuss the application of hydro-shearing as a geothermal stimulation technique. The possibility that silica prop pant may dissolve in the aggressive high-temperature fluid environment of some geothermal reservoirs whereas slip on rough fractures develops aperture increase without the need for prop pant is also presented as an argument in favor of hydro shearing.

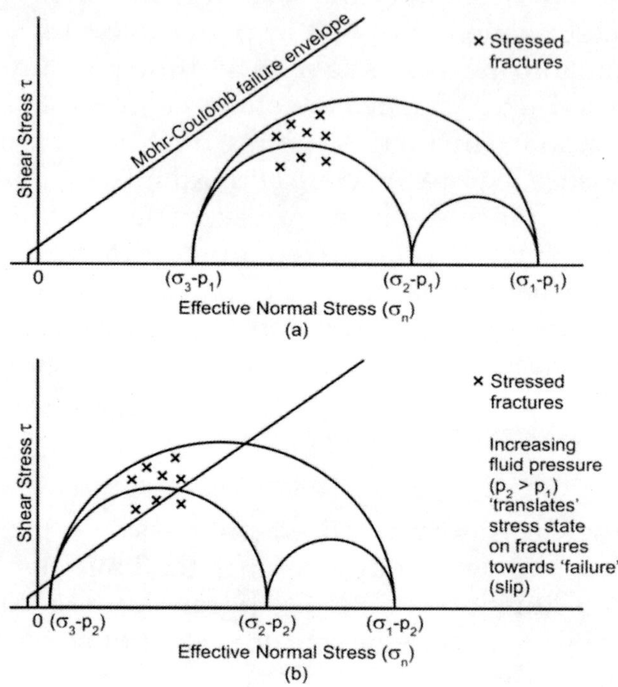

FIGURE 4. Hydro-shearing – a procedure to generate slip on pre-existing fractures by increasing the fluid pressure to a level below that required to generate a hydraulic fracture.

DEFORMATION AND FAILURE OF ROCK IN SITU

As with fabricated materials, the deformation and failure of brittle rock is also dependent strongly on fractures and discontinuities. In a rock mass, however, the fractures occur over a very wide range of scales from sub-microscopic to the size of tectonic plates. A large specimen of rock will probably include some large fractures, and as the scale of the rock mass increases, fractures from different tectonic epochs.

Study of fracture systems underground in mines and in civil engineering projects allow systems of fractures to be identified and classified statistically into discrete fracture networks (DFN's). The network will include intersecting sets of planar fractures, but individual fractures will tend to be of different lengths, and though organized in two or three spatial orientations, of variable, finite length and not collinear.

Figure 7 presents a two-dimensional illustration of the application of DFN's to the numerical modeling of a fractured rock mass. The in-situ rock mass is considered as a large specimen of intact rock that has been transected by the DFN determined from field observations and fracture mapping underground or at surface outcrops. The properties of the intact rock are built into a Bonded Particle Model of the rock (using the Particle Flow Code (PFC) code) based on results of laboratory tests of the intact rock deformability and strength. The intact rock representation is shown on the left of Figure 6. The DFN (shown on the upper right in Figure 6) then is superimposed onto the intact rock.

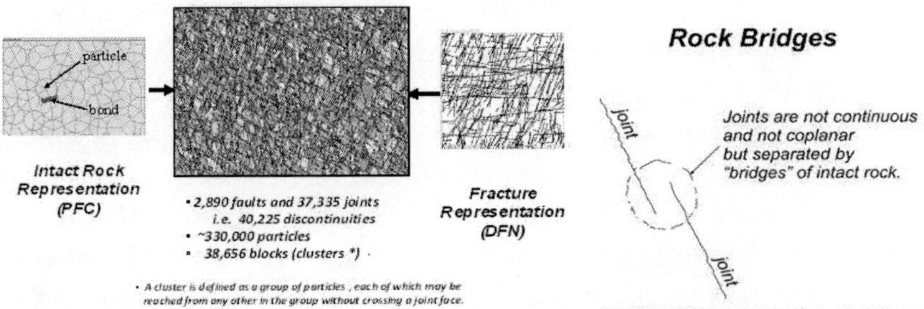

FIGURE 5. The Synthetic Rock Mass (SRM) representation of a fractured rock mass (in two dimensions). Damjanac et al. (2013) present a discussion of the 'construction' of an SRM in three dimensions. Pierce (2011) presents a comprehensive discussion of practical guidelines and factors involved in the construction of DFN's.

Cohesion and friction values are assigned to the joint planes.[10] - The 'unconfined' strength of a typical large SRM is of the order of a few percent of an intact rock specimen of the same rock (Cundall, 2008). Much of the in-situ strength is derived, of course, from the in-situ stresses imposed on the SRM in situ. One of the consequences of the finite length and lack of collinearity of joint sets in DFN's is the formation of bridges of intact rock Figure 4 within the SRM. These bridges provide regions of intact rock, and of stress concentration, in the SRM and account for a significant part of the overall strength of the rock mass. Earlier models of a rock mass, considered to consist of several sets of through-going fractures, exhibited much lower rock mass strength (Hoek and Brown, 1980).

Figure 5 presents selected extracts from a two-dimensional PFC simulation of the development of a hydraulic fracture in a jointed Synthetic Rock Mass. The SRM model was developed following the procedure outlined in Figure 5. The joint distribution was based on a DFN obtained at the Northparkes Mine in Australia.[11] - Figure 5(a) shows the location of a vertical borehole that was pressurized by fluid until a hydraulic fracture was initiated. The rock mass is assumed to be impermeable. (The path of the fracture has been traced in blue for clarity.) Displacements in the rock mass produced by the hydraulic fracture are shown as vectors on each side of the fracture. It is seen that the fracture started more or less symmetrically on each side of the borehole, but propagation of the right wing was arrested when the hydraulic fracture encountered an adversely oriented pre-existing joint (Figure 5(b)). With increasing pressure, in the borehole, the hydraulic fracture continued to extend asymmetrically towards the left (Figures 5(c) and 5(d) Figure 5(d) is simply an enlarged view of Figure. It is seen that the propagating fracture extended partially by opening existing fractures and partially by developing new fractures through intact rock. Although local deviations occur, the overall path of fracture growth is approximately perpendicular to the direction of the minimum compression stress. The existing fractures introduce an asymmetry to the rock mass. In terms of the idealized symmetric crack of Figure 2, the system in Figure 3can be considered as two cracks, one extending to the right and one to the left of the borehole with a higher 'fracture toughness' on the right compared to the left, etc.

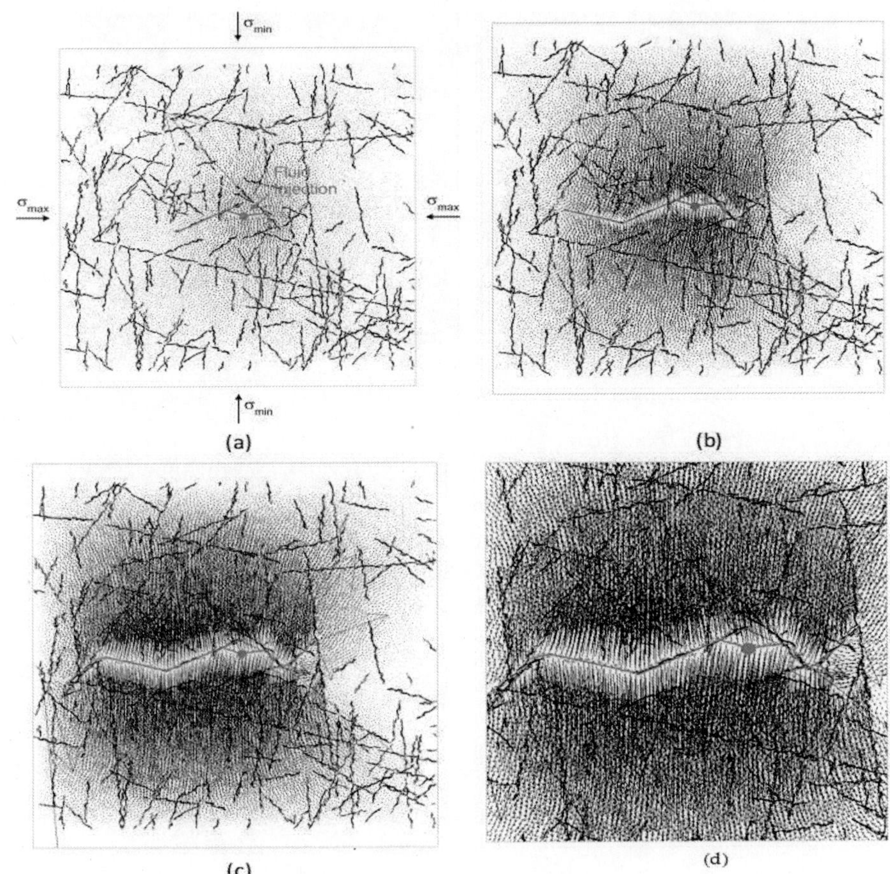

FIGURE 6. Extracts from simulation of the propagation of a hydraulic fracture in a two-dimensional impermeable SRM (Synthetic Rock Mass). (The horizontal stress σ_{max} is 29 MPa and the vertical stress σ_{min} is 12 MPa – Figure 5(a)). Note that the intact rock between the fractures has a finite strength and can break by rupture of the cemented bonded particles shown in Figure. The pressure required to propagate the fracture after breakdown was approximately 10 MPa above the minimum (i.e., least compressive) principal.

Jeffrey et al. (2009) conducted an underground test in the Northparkes Mine, Australia to observe the propagation of a hydraulic fracture in naturally fractured tock. Figure 7 shows part of the path of the fracture, as seen in a tunnel excavated into the fractured rock. The fracture path shows similar characteristics to those shown in the PFC simulation in Figure 6.

FIGURE 7. Hydraulic fracture (green plastic) crossing a shear zone on the face of a tunnel excavated through the fracture. "The arrows indicate the trace of the fracture with green plastic contained in it. There is no clear fracture between points 1 and 2 but the fracture may have crossed this zone either deeper into the rock or in the rock that has been excavated. Approximately 2 m of fracture extent is visible" (Jeffrey et al., 2009).

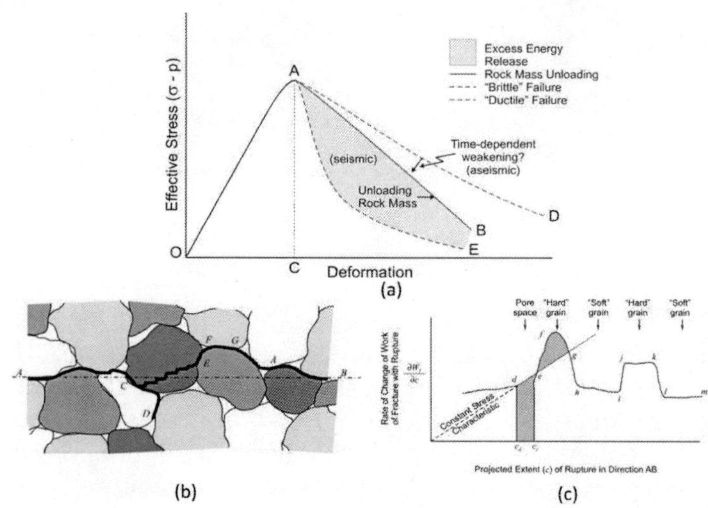

FIGURE 8. Energy changes during propagation of a fracture through heterogeneous rock.

The energy required to initiate crack propagation is represented by the area OAC in Figure 7(a). Whether or not the crack will extend depends on the energy that becomes available from the intact rock around the crack. If the energy released from the rock mass, represented by the area under the red curve AB, is greater than the energy required to extend the crack, represented by the area under curve AE, then the crack will extend; the excess energy represented by the shaded area serves to accelerate the crack and release seismic energy. If the energy required to extend the crack is represented by the area under the green curve AD, it is greater than the energy that would be released from the rock mass, and hence the crack would not extend. It is possible that the crack could exhibit some form of time-dependent weakening (e.g., due to fluid flow to the crack, viscous behavior, etc.) such that the energy required to extend the crack would be reduced. This could lead to crack extension, i.e., as the slope AD increased to overlap AB, but with no excess energy to produce seismicity. Figures 7(b) and 7(c)[12] - illustrate another feature of crack extension on the granular scale. The energy required to extend a crack through or around a grain will be variable; the fracture may encounter pore spaces where no crack energy is required. Application of a constant load to such a heterogeneous system will result in local acceleration and deceleration of the crack-producing bursts of micro seismicity. Similar effects can arise in rock fracture propagation at all scales.

It is worth noting that all of these processes of fracture propagation, albeit complex, develop in accordance with the principle of seeking the minimum potential energy of the system.

Much of the preceding discussion has focused on two-dimensional analysis or models. In reality, we are dealing with three- dimensional space (as noted in Figure 6), plus the influence of time (e.g., with respect to fluid flow, or time-dependent rock properties). Figure 8 provides an example from an actual record of hydraulic fracture propagation.

Figure 8 shows the sequence of micro seismic events observed during hydraulic fracture stimulation ('treatment' in Figure 8(a)) of a borehole. Early time events are shown as green dots; later events are in red. The micro seismic pattern indicates that fracturing started on both sides of the borehole at the injection horizon, but then moved up some 100 m to a higher horizon. As pumping continued, fracturing continued (red locations) on both horizons. It was concluded that the initial fracture in the lower horizon had intercepted a high-angle fault, allowing injection fluid to move to the higher level where it opened up and extended another fracture. Continued pumping led to fracture extension on both horizons.

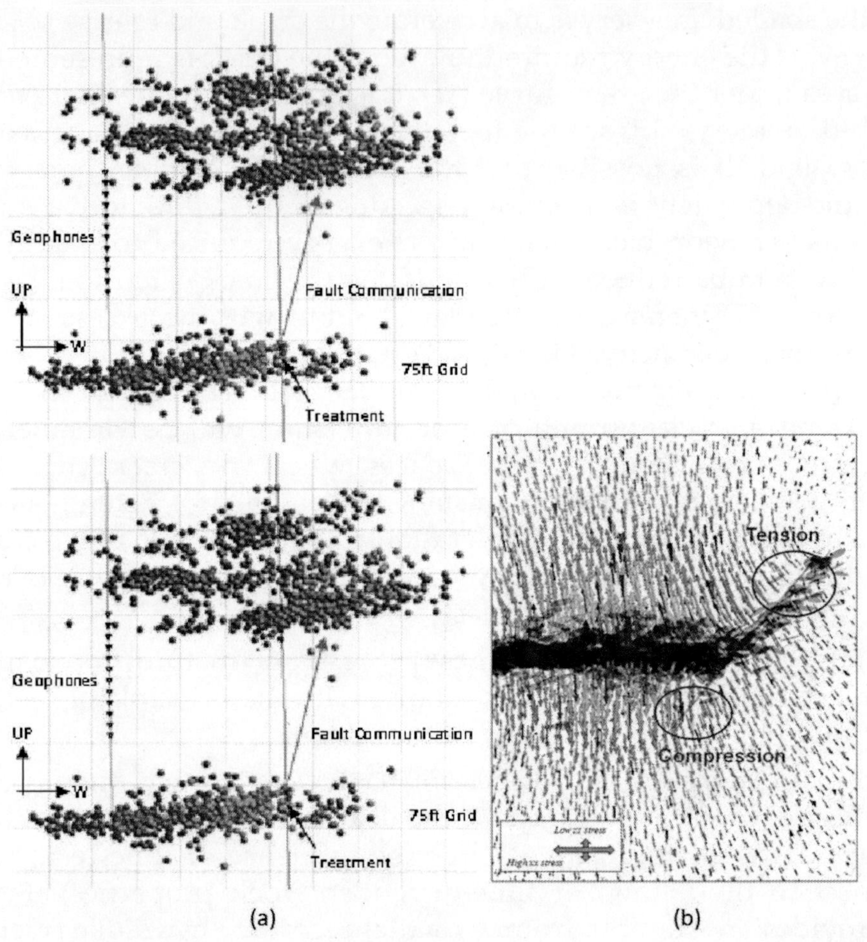

FIGURE 9. (a) Micro seismicity observed during hydraulic fracturing in a deep borehole; (b) numerical 'explanation' of the behavior observed in (a).

Numerical analysis Figure indicated that initial fracture propagation at the lower level resulted in induced tension on the fault above the horizon, but compression on the fault below the lower injection horizon. This explains why injection fluid did not penetrate along the fault below the horizon, and provides a good illustration of the benefit of combining numerical analysis with field observation in understanding fracturing processes.

MICROSEISMICITY AS AN INDICATOR OF SLIP ON FRACTURES

Micro seismicity stimulated during hydraulic fracturing and associated stimulation techniques (e.g., hydro shear) are often used to indicate slip and deformation on fractures in the rock. In some cases, it is tacitly assumed that absence of micro seismicity indicates absence of slip or deformation. In fact, there is growing evidence that micro seismicity does not present a complete picture of deformations induced by stimulation or other effects leading to stress change. Figure 9, reproduced from Cornet (2012) (with permission from the author), shows P-wave velocity changes observed by 4D (time-dependent) tomography during the stimulation of the borehole GPK2 in the year 2000. A detailed discussion of the procedure used to observe and determine the P-wave changes is presented by Calo et al. (2012).

It is seen that the region of detected micro seismicity (the cloud of black dots is small compared to the region where the P-wave velocity is reduced by as much as 20% in some regions). Some of the changes in velocity were temporary, suggesting that they may be related to temporal changes in fluid pressure; other changes appeared to be more permanent deformation that occurred a seismically.

These observations indicate that micro seismicity, although a valuable indicator of the response of a rock mass to stimulation by fluid injection, does not identify the complete region influenced by a stimulation.

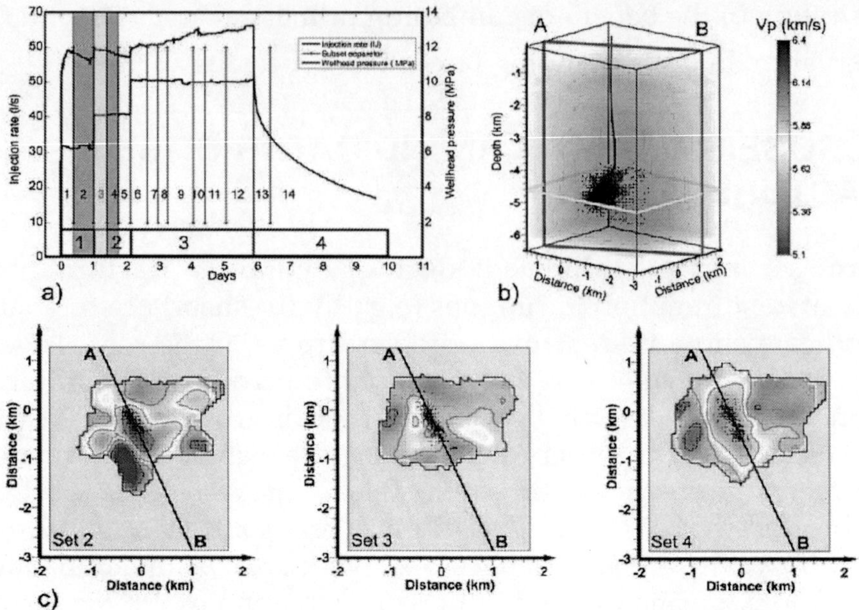

FIGURE 10. Aseismic slip induced by forced fluid flow as detected by P-wave tomography. (Soultz- sous- Fôrets, France. (a) The injection program (black curve is flow rate, blue curve is well head pressure, horizontal axis is time in days); (b) 3D view of the seismic cloud with respect to the GPK2 borehole. Vertical axis is depth and horizontal axes are distances respectively toward the north and toward the east; and (c) horizontal projections corresponding to the yellow horizontal plane. The vertical green plane is shown as line AB in the plots of part c. P-wave velocity tomography for sets 2, 3 and 4 are indicated respectively by orange, yellow and green colors in the injection program. The vertical axis corresponds to North.

IN-SITU STRESS

As already noted, hydraulic fractures tend to develop in a more or less planar fashion, extending normal to the minimum regional principal stress. Determining the direction, and perhaps the magnitude, of the regional minimum stress is an important element of hydraulic fracturing strategy, especially with the development of directional drilling, which allows borehole to be drilled in the direction considered most favorable for fracturing with respect to stress direction. (see e.g., Figure 15 and related discussion).

Determination of the in-situ stress state also can be a significant challenge.

Stress in rock is distributed throughout the mass, and is influenced by the complicated structure of the mass [13] - . Most techniques of stress determination rely on what are essentially 'point' determinations. One difficulty of determining the regional stress is illustrated by the simple, albeit somewhat artificial, example of Figure 11. This shows a two-dimensional numerical model of the stress distribution in an elastic plate containing several finite frictional fractures.

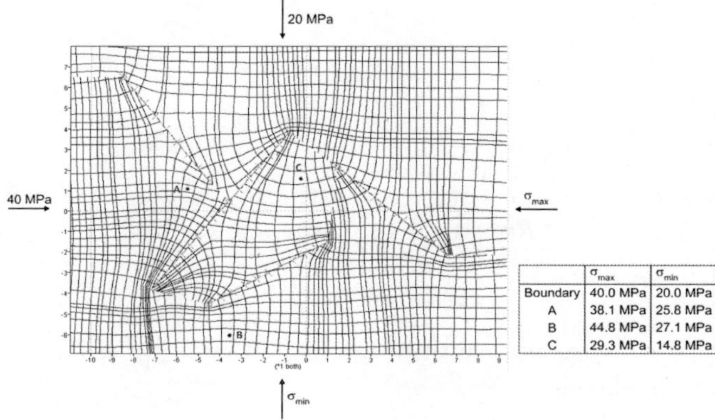

	σ_{max}	σ_{min}
Boundary	40.0 MPa	20.0 MPa
A	38.1 MPa	25.8 MPa
B	44.8 MPa	27.1 MPa
C	29.3 MPa	14.8 MPa

FIGURE 11.Influence of frictional cracks on the distribution and orientation of principal stresses, illustrative example.

The exercise serves to illustrate the difficulty of making stress determinations from local point measurements, be they in a borehole or on the surface. Stresses can change in orientation and magnitude locally due to geological in homogeneities, fractures, faults, etc., many of which may be hidden or cannot be observed from the measurement location. Although determinations made at points A and B are reasonably close to the boundary values, point C is considerably different, and the directions of principal stress, as indicated by the principal stress trajectories, can be very different from the (regional) orientations, i.e., at the model boundary.

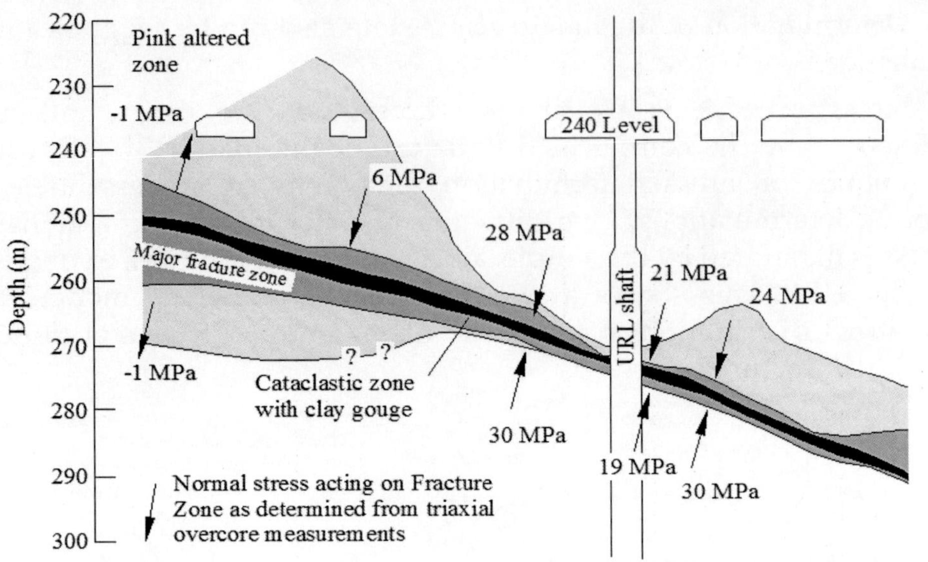

Observed Variability of Normal Stress Across a Thrust Fault Underground Research Laboratory Pinawa, Canada.

FIGURE 12. Normal stress variation across a thrust fault, Underground Research Laboratory, Canada.

Figure 12 provides an actual example of the variability of stress over relatively short distances. (The vertical and horizontal scales are equal in Figure 12). In this case, the main interest was to assess how normal stresses were affected by the thickness of gouge in the plane of the thrust fault.

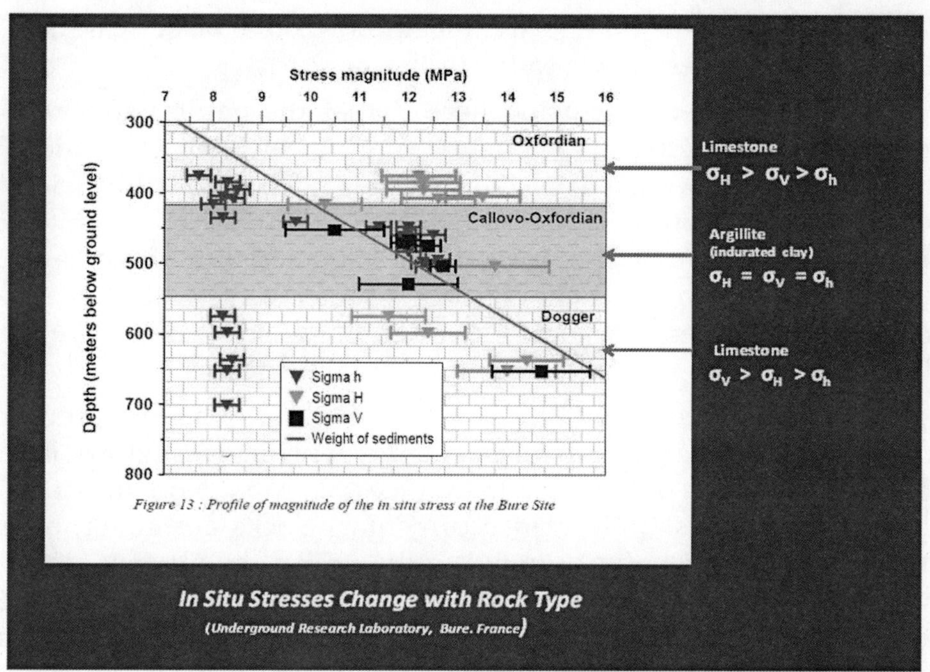

FIGURE 13. Observed stress distributions in argillite and limestone's at the Underground Research Laboratory, Bure, France.

Figure 13 illustrates another important geological influence on stress distribution, changing lithology. This example is from the French Underground Research Laboratory (URL) [14] - at Bure in NE France. Laboratory tests on specimens of the Callovo-Oxfordien Argillite indicate a long-term viscosity of this rock suggesting that any imposed deviatoric stresses would tend towards an isotropic stress state over the order of 10 million years.

Test specimens from the limestone's above and below the argillite do not appear to exhibit such viscosity. The stress distributions determined from field measurements support such differences in rheological characteristics of the rock formations.

Commenting on the in-situ stresses observations at Bure (i.e., as shown in Figure 13) Cornet (2012) notes as follows:

"Further, the complete absence of microseismicity in the Paris Basin (Grünthal and Wahlström, 2003, Fig. 4) and the absence of large scale horizontal motion as detected by GPS monitoring (Nocquet and

Calais, 2004) indicate that no significant horizontal large-scale active deformation process exists today in this area.

"The important conclusion here is that the natural stress field measured on a 100 km^2 area at depth ranging between 300 m and 700 m does not vary linearly with depth and is not controlled by friction on preexisting well- oriented faults. Rather, the stress magnitudes seem to be controlled by the creeping characteristics of the various layers rather than by their elastic characteristics, with a loading mechanism that remains to be identified but which is neither related directly to gravity nor apparently to present tectonics.

"It is concluded here that the smoothing out of stress variations with depth into linear trends may be convenient for gross extrapolation to greater depth. But it should not be taken as a demonstration that vertical stress profiles in sedimentary rocks are governed by friction along optimally oriented faults, given the absence of both micro seismicity and actively creeping fault. It should not be used for integrating together stress tensor components obtained within layers with different rheological characteristics."

Other examples could be cited, but the message is clear. Determination of in-situ stress in rock is an extremely challenging task, with results subject to considerable variability and uncertainty.

Stress orientations can be estimated from consideration of regional tectonics, faulting and interpretation of evidence from local structural geology supported in some cases by evidence based on borehole logs (e.g., tensile fractures induced along the well bore). Stress magnitudes are, in general, more difficult to determine and usually less significant, except as indicators of how stresses may be distributed across a site where the geology and engineering design are complex. In such cases, interpretation of stress distribution is best done in conjunction with a numerical model of the site, preferably one that includes the influence of important uncertainties and discussion with structural geologists familiar with the area under study.

'CRITICAL STRESS STATE' IN THE EARTH'S CRUST

It is sometimes asserted that the Earth's crust is everywhere close to a 'critical state of stress,' i.e., that a small change in the devatoric stress in the rock is likely to produce slip on one or more faults with associated seismic activity. The current global interest in development of major resources of natural gas, the central role of hydraulic fracturing in this development, and the public apprehension that hydraulic fracturing will 'trigger earthquakes' has led to strong opposition to fracturing, and even legislation to ban the use of hydraulic fracturing in some countries and some States in the USA.

As illustrated by Figure 14, the seismic hazard, (i.e., probability of a damaging earthquake) varies very considerably from place to place. Thus, an earthquake of a given magnitude is 1000 times more likely to occur in Southern California than it is in the Eastern United States. The hazard is even lower in regions such as Texas, North Dakota and in the stable Canadian Shield region of the North American tectonic plate.

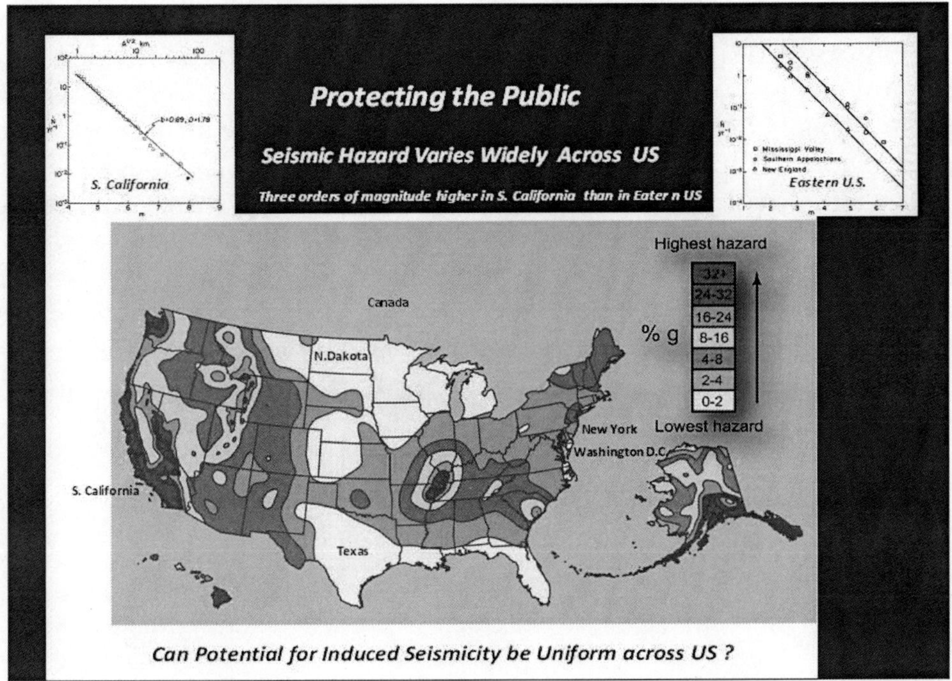

FIGURE 14. Seismic hazard map of the United States — US Geological Survey.

While many earthquakes are initiated at depths considerably greater than depths where hydraulic fracturing is applied, it seems plausible to suggest that there may be less potential for fracturing to induce seismic activity in regions that have low seismic hazard. Also, as indicated by the comments of Cornet in the previous section of this paper, there is evidence that the critical stress hypothesis warrants detailed scrutiny, at least. This could have major implications for development of the world's major natural gas and EGS (enhanced geothermal systems) resources. Two recent studies, National Research Council (2012) and Royal Society - Royal Academy of Engineering (2012), have each concluded that the risk that hydraulic fracturing as used in development of energy resources would trigger significant seismic activity is small, but it would be valuable to examine the critical stress hypothesis more rigorously than has been done to date.

HYDRAULIC FRACTURING IN TIGHT SHALES

The development of inclined and horizontal drilling (see Appendix 1 - Figure A1-2) has helped stimulate intense activity to develop natural gas production from so-called tight shale, i.e., rock in which natural gas is held tightly within the very fine pore structure of the rock. Figure 15 illustrates the procedure used to stimulate these shale's. The well is drilled horizontally in the gas-bearing formation, more or less in the direction of the minimum principal stress. Hydraulic fractures are generated (and propped) at intervals along the well to generate a network of connected flow paths that will allow the gas to flow to the well. Depth (i.e., extent) and spacing of the fractures should be optimized to produce the formations effectively. Bunger) discuss the factors in the design of an effective fracture strategy.

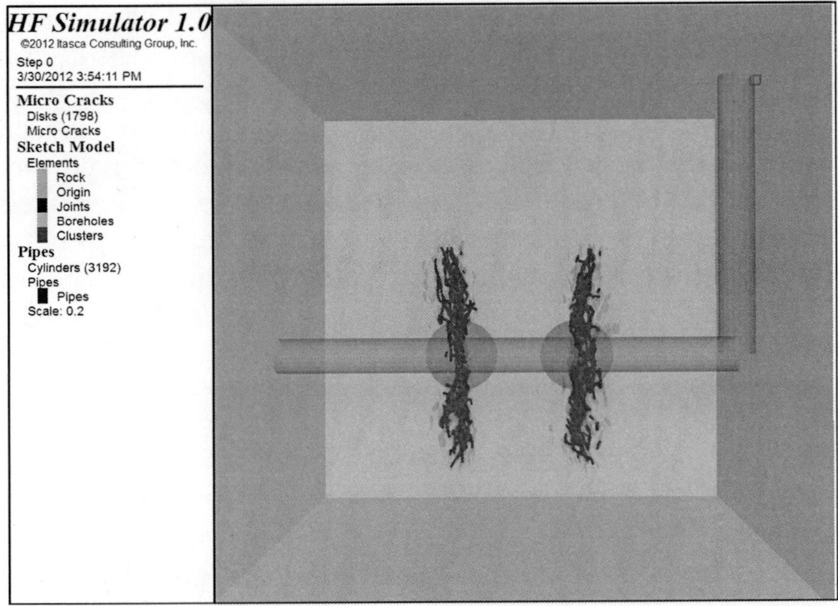

FIGURE 15. Staged hydraulic fracturing in a horizontal well. There may be many such wells along the horizontal well.

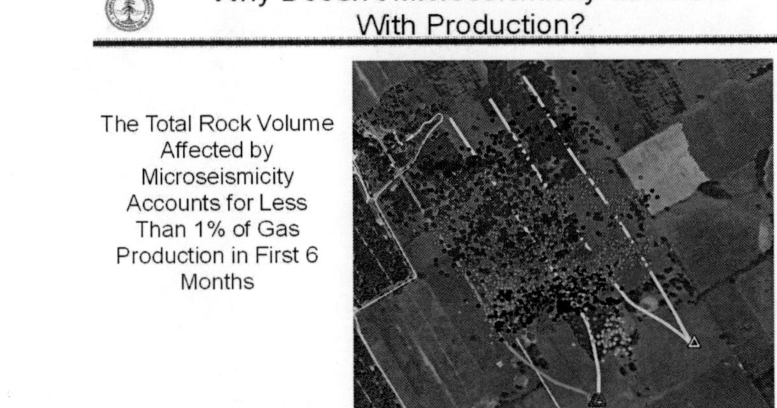

FIGURE 16. The volume of rock defined by microseismicity is a very small fraction of the volume producing gas.

Figure 16 shows a slide from a recent presentation by Prof. Mark Zoback, who kindly agreed to allow the author to include it here. Although on a somewhat smaller scale, the fact that considerable deformation and fracturing must be taking place that is not associated with detected microseismicity is similar to the phenomena discussed in connection with Figure 10. Prof. Zoback refers to such aseismic deformation as slow slip, and is conducting research to understand the underlying mechanisms, including the possible influence of the clay content of the shale. As can be seen in Figure 17 (courtesy of Prof. Zoback), the clay content can be large.

Average Shale Properties

	BARNETT	MARCELLUS	EAGLE FORD	FLOYD
Depth (ft)	3 - 9,000	2 - 9,500	4 - 13,500	6 - 13,000
TOC (%)	1 - 10	1 - 15	2 - 7	1 - 7
RO (%)	0.7 - 2.3	0.5 - 4+	0.5 - 1.7	0.7 - 2+
Porosity (%)	2 - 14	2 - 15	6 - 14	1 - 12
Qtz + Calcite (%)	40 - 50	40 - 60	50 - 80	20 - 30
Clay (%)	20 - 40	30 - 50	15 - 35	45 - 65
Areal Extent (mi²)	22,000	60,000	15,000	6,000
Resource Size (Tcf)	25 - 250	50 - 500	10 - 100	<<1

How many Floyd Shales are There?

33

FIGURE 17. Clay content of some typical 'tight' gas shales.

Figure 18 illustrates the very fine, micron scale, pore structure of typical tight shale. Although the mechanism(s) by which flow pathways are established in such a fine structure is not clear, the level of micro seismic energy release associated with brittle breakage of one or a few bonds will be very small and of high frequency (such that the radiated energy would be rapidly attenuated), and hence, not detectable by any geophone. Thus, absence of micro seismicity may not indicate an absence of breakage of brittle bonds. Some mechanism must be operative that generates flow pathways. Intuitively, it might be expected that the clay content of the shale might lead to ductile and viscous deformation that could tend to close the pathways.

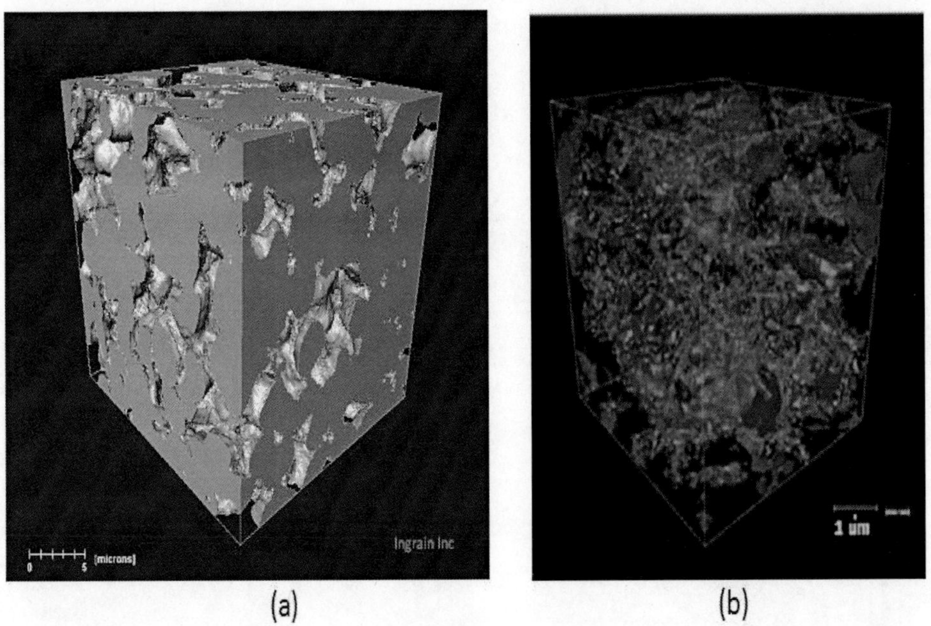

FIGURE 18. (a) Outer surface of a FIB-SEM (Focused Ion Beam- Scanning Electron Microscope) volume of Eagle Ford Shale; (b) Transparency view of the distribution of connected pores (blue), isolated pores (red) and organic matter (green). (Courtesy of Prof. Amos Nur and J. Wallis (see Wallis et al., (2012) for details of technology.)

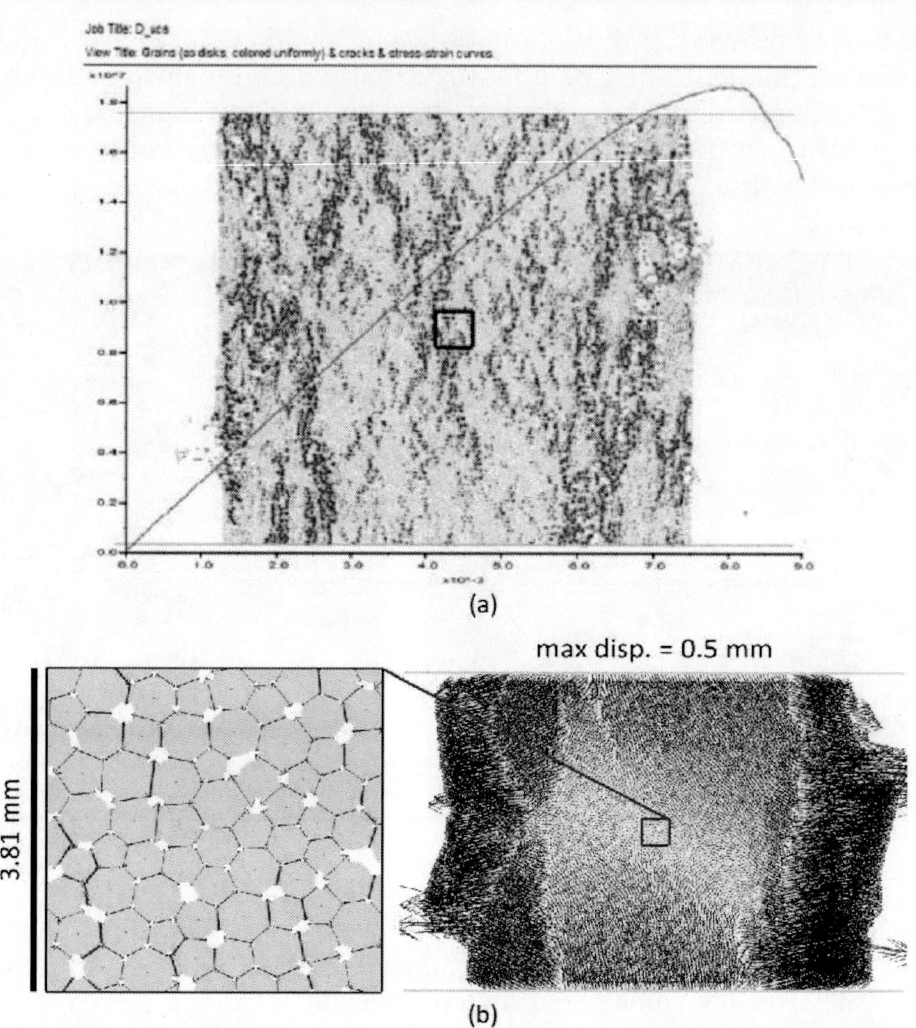

FIGURE 19. Micro-rupture of bonds within a *PFC* model of a rock loaded to failure, and beyond, in uniaxial compression. The darker red regions in (a) indicate coalescence of smaller groups of bonds that have ruptured. Eventually these larger regions develop to provide a mechanism that leads to collapse of the specimen. It is seen that bond breakage occurs throughout the specimen as the load is increased. The larger dark red regions will release larger amplitude, lower frequency waves that can be detected, whereas the smaller 'pathways' cannot be detected seismically. The load-deformation curve is shown as an 'overlay' on the specimen.

FRACTURE NETWORK ENGINEERING

This paper has emphasized the central role of fractures in rock, primarily natural fractures developed on a wide spectrum of scales over many tectonic epochs and many millions of years. These fractures and fracture systems are of special significance with respect to hydraulic fracturing and related techniques of fluid injection into rock since the fluid will tend to seek out those fractures that can be more readily opened against the local in-situ stress field as the fluid is injected. Given the complexity and lack of information on the fracture system, stress environment, etc., how can the engineering of hydraulic fracturing and related fluid injection programs advance most effectively?

Confronted with the same complexity of rock in situ, civil engineers and mining engineers have tended to adopt the 'Observational Approach' (Peck, 1969). In essence, this approach involves developing an initial engineering design for the problem, based on a first assessment/estimate of the rock (or soil) properties. Observe the actual performance and modify the initial design as needed to arrive at the desired performance. An example of the Observational Approach (as used in the New Austrian Tunneling Method) is discussed inFairhurst and Carranza-Torres (2002), see pp. 24-30.

Application of the Observational Approach to Hydraulic Fracturing and related fluid injection techniques faces some disadvantages and some advantages. We do not have 3D access to the engineering site. We do have powerful numerical modeling tools to help make a more informed initial estimate of how the system will perform; and we have sensing systems, both down hole and remote. Figure 20 illustrates a procedure that tries to apply the Observational Approach to hydraulic fracturing and related systems. The illustration describes an application to the extraction of Geothermal Energy.

Stones have begun to speak, because an ear is there to hear them. .....

Cloos, Conversations with the Earth (1954), 4

Microseismicity –predicted and observed.

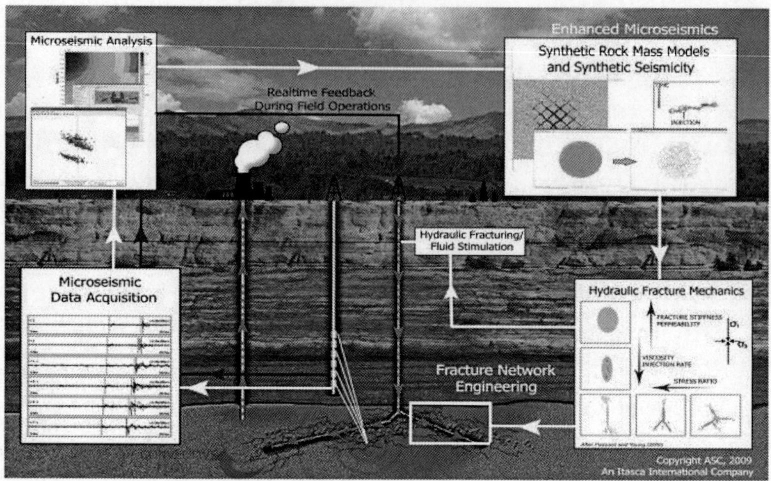

Fracture Network Engineering. *Synthetic Rock Mass and Synthetic Seismicity Models are compared with observed microseismic signals for* ***real time control of fracture network development.*** *(Enhanced Geothermal Systems.)*

FIGURE 20. Fracture network engineering system.

In this application, an initial design approach is developed based on a numerical modeling study incorporating any available data, insight, etc., on the site. This model provides an initial prediction of the performance. Instrumentation, both down hole and on-surface observes the initial response of the system and compares it with the prediction. This triggers a feedback signal to modify the design input to move the performance closer to the one desired. This iteration continues, changing progressively towards the performance desired.

Although the writer knows of no such Fracture Network Engineering system currently in operation, many of the components are available and it is time to start.

CONCLUSIONS

Expectations for higher living standards of a rising world population, and the associated demand for Earth's resources of energy, minerals and water, lead inevitably to greater focus on resources of the subsurface.

This focus includes the need to develop improved technology to develop these resources, and a better understanding of the nature of the subsurface environment as an engineering material.

Earthquakes and dynamic releases of energy are a daily reminder that on the global scale, Earth is critically stressed, and constantly trying to adjust seeking to achieve a condition of minimum potential energy for the entire system.

Ongoing for many, many millions of years, such adjustments have resulted in the heterogeneous assembly of blocks of rock bounded by essentially planar surfaces; fault, fractures and similar 'discontinuities' varying in scale from tectonic plates and continents down to micron and even nanometers.

Some of these volumes are critically stressed; others are far from a critical condition. National maps of seismic hazards provide evidence of this heterogeneity on a larger scale.

Although Earth Resource Engineering activities may be kilometers in extent, they are small-scale within the larger Earth context. Subsurface engineering in a critically stressed region can be a much different challenge than in a stable region. It is important to assess the initial conditions carefully for each case, and especially where fluid injection is a main component of a project.

The sub-surface is opaque in several ways. Details of the key features that can control the response to an engineering activity in the sub-surface are often unknown. Problems are data-limited. This is particularly the case when the engineering is based on deep borehole systems, as in hydraulic fracturing and related fluid injection technologies.

Although operating in ways that may appear complex, the response of the subsurface to stimulation does obey the laws of Newtonian mechanics, and it is clear that pre-existing natural discontinuities have a major influence on how the subsurface responds to engineered changes.

The advent of powerful computers and developments in numerical modeling provide a potentially major tool to help develop better-informed strategies of subsurface engineering. Used interactively in close conjunction with instrumentation, both down hole and surface based, it should be possible to progressively develop a mechanics-informed understanding and path forward for more effective subsurface engineering.

Much as the field of Fracture Mechanics has led, and continues to lead, to major technological improvements for fabricated materials, so can development of the field of Rock Fracture Mechanics be of transformative value to subsurface engineering, and to society in general.

Hydraulic fracturing and related injection-stimulation systems will certainly be a central element in the future of Earth Resource Engineering. The organizers of HF 2013 are to be commended for focusing attention on this critically important topic.

APPENDIX 1

Earth Resources Engineering

In 2006, the US Academy of Engineering introduced the term 'Earth Resources Engineering' to replace 'Petroleum, Mining and Geological Engineering' in recognition of the broader range of engineering activities and concerns associated with use of the subsurface. The new title, it is hoped, will also stimulate important synergies between the various disciplines involved. Mining and civil engineers, for example, have direct three-dimensional access to the subsurface not available to colleagues in other subsurface activities. This access provides a major opportunity to conduct research and gain understanding of the mechanics of subsurface processes under actual in-situ conditions, as exemplified by Jeffrey et al. (2009), see Figure A1-1.

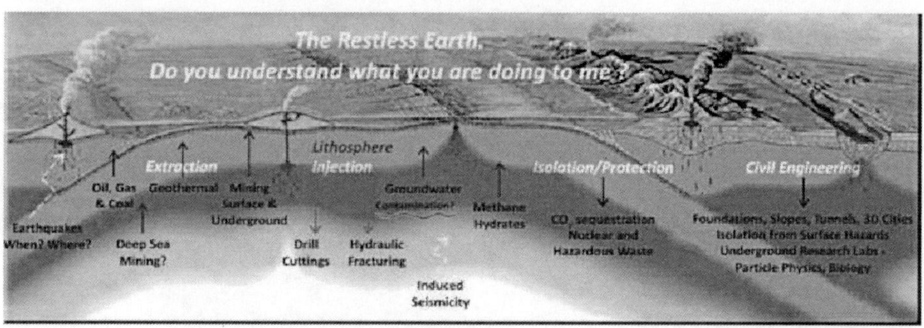

FIGURE A1-1.The restless Earth. Earth Resource Engineering activities are all confined to a very shallow part of the 40 km -700 km thick Earth's solid crust (lithosphere). Deepest borehole ~ 12 km; mine ~ 4km. Rock stress increases vertically σv ~ 27MPa/km; laterally σh~ (0.5- 3.0).σv: Pore water pressure p = 10 MPa /km; temperature increase ~25°C /km depth.

Study of slip on active faults is a good example.

"The physics of earthquake processes has remained enigmatic due partly to a lack of direct and near-field observations that are essential for the validation of models and concepts. DAFSAM [15] - proposes to reduce significantly this limitation by conducting research in deep mines that are unique laboratories for full-scale analysis of seismogenic processes. The mines provide a 'missing link' that bridges between the failure of simple and small samples in laboratory experiments, and earthquakes along complex and large faults in the crust. There is no practical way to conduct such analyses in other environment. To unravel the complexity of earthquake processes, this project is designed as integrated multidisciplinary studies of specialists from seismology, structural geology, mining and rock engineering, geophysics, rock mechanics, geochemistry and geobiology. The scientific objectives of the project are the characterization of near-field behavior of active faults before, during and after earthquakes".[16] - See also http://www.iris.edu/hq/instrumentation_meeting/files/pdfs/IRIS_Johnston.pdf

Petroleum engineers can now reach depths in excess of 6 km and have developed advanced drilling control technologies that allow precise access to locations extending horizontally to more than 10-15 km from a single vertical hole (see Figure 2).

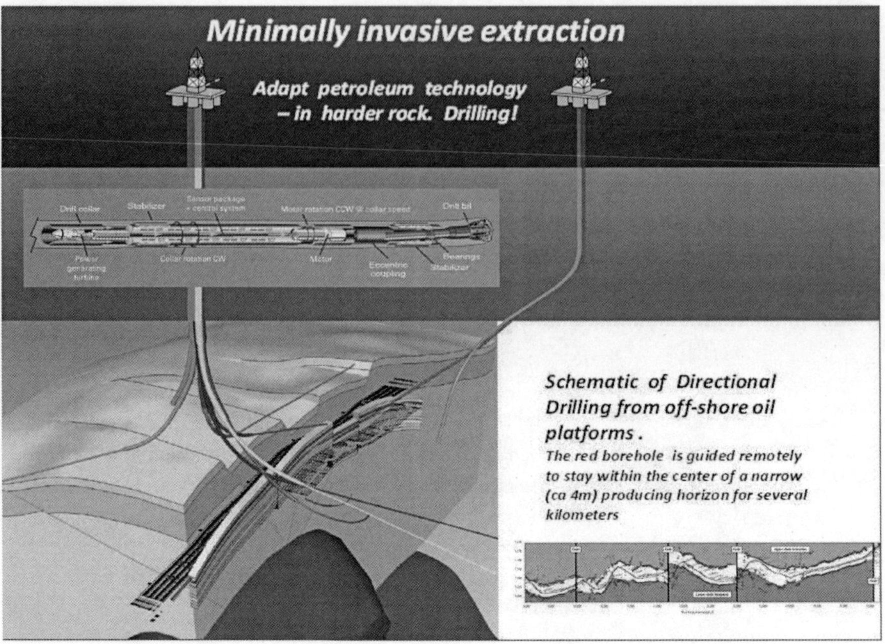

FIGURE A1-2. Schematic illustration of directional drilling for petroleum production.

These and related developments are stimulating interest in application of borehole technologies to other areas of subsurface engineering, including the development of less-invasive mining technologies, i.e., borehole extraction of minerals. Some applications, e.g., where crystalline rocks are involved, are contingent on the development of significantly lower-cost drilling technologies. The critical dependence of society on reliable and economic subsurface engineering is illustrated by the fact that currently more than 60% of the world's energy is delivered via a borehole. The Deep-water Horizon accident in the Gulf of Mexico in April 2010 provides a sober example of the consequences of error. In summary, hydraulic fracturing and related stimulation technologies are likely to see application to an increasing range of subsurface engineering challenges. HF2013, the first International Conference for Effective and Sustainable Hydraulic Fracturing, is very timely.

APPENDIX 2

Effect of Coring In Pre-Stressed Rock

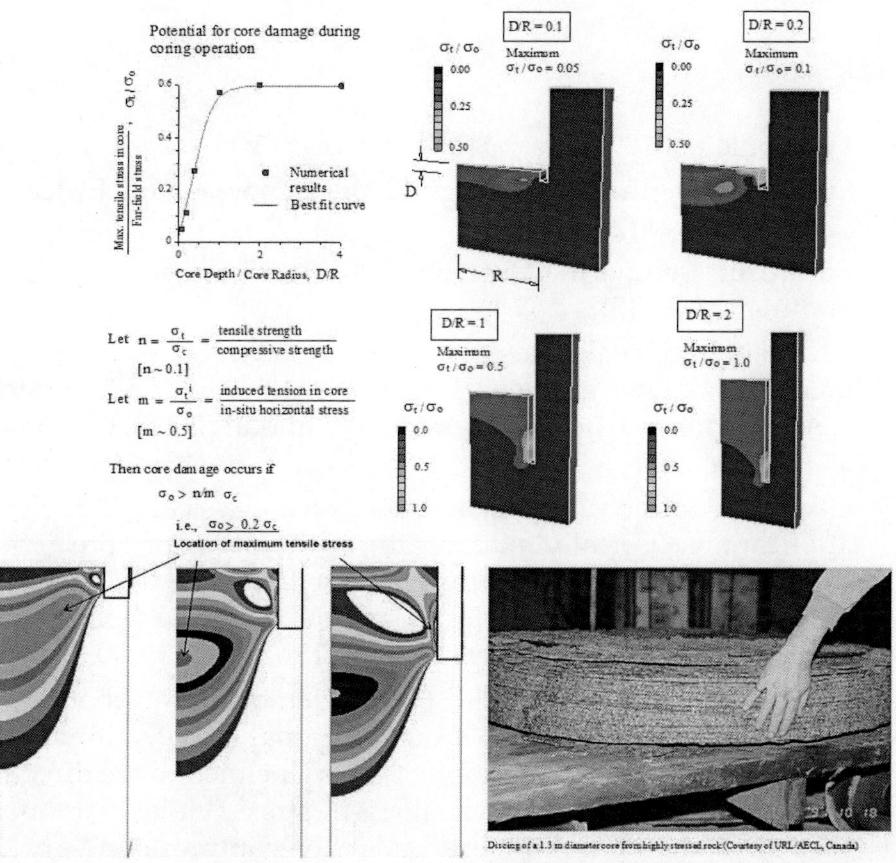

FIGURE A-2.1. Tensile stress concentrations induced in a brittle rock during coring.

The consequences of disturbing a pre-stressed rock medium are illustrated by examining the rock coring operation. Figure A2-1 shows the stress concentrations in a rock core in a brittle rock. If the in-situ stress normal to the axis of drilling is sufficiently high tensile cracks can develop in the core. Where lateral stresses are very high, then tensile 'spalling' may result, as shown in the photograph of the bottom right of Figure A2-1. Where the rock is more 'ductile' the core may undergo permanent deformation without fracturing. In both cases, the mechanical properties of these cores may differ significantly from those of the rock in situ from which the core was obtained.

Notes

[1] 1 exajoule =1018 joules = 1018 watt.seconds.

[2] Future of Geothermal Energy (2005) Synopsis and Executive Summaryp.1-4 (2).

[3] Future of Geothermal Energy (2005)Synopsis and Executive Summaryp.1-5 (5).

[4] A fractured rock mass is typically about two orders of magnitude lower in strength than the strength of a laboratory specimen taken from the rock mass [Cundall (2008); Cundall et al, (2008)].

[5] Hydraulic fractures generated in classical petroleum applications typically extend (2b) of the order of 25m ~ 50m from a wellbore. The fracture aperture (2a) at the wellbore then will be typically of the order of 0.01 m. Thus, the tensile stress concentration at the tip is very high of the order of 103.

[6] In his second paper, Griffith (1924), demonstrated that tensile stresses also developed around similar cracks loaded in compression, provided the cracks were inclined to the direction of the major principal (compressive) stress.(He also assumed that the cracks did not close under the compression.) For the optimum crack inclination, an applied compressive stress of eight times the magnitude of the tensile strength was required to develop a tensile stress on the crack boundary (close to, but not at the apex of the crack) equal to the limiting value in the tensile test. He concluded that the uniaxial compressive strength of a brittle material should be eight times greater than the tensile strength. Interestingly, he did not invoke his second (minimum potential energy) criterion. It was later determined that although a tensile crack could initiate in a compressive stress regime as predicted by Griffith (1924), the crack was

stable (i.e., did not satisfy the minimum potential energy criterion). The compressive/tensile strength ratio is greater than 8 (see Hoek and Bieniawski, 1966).

[7] Tension is assumed to be positive in Figure 3.

[8] A typical hydraulic fracture may have a length (2a) of the order of 50m and a maximum aperture (2b) of 5mm, so that the stress concentration will be of the order of 2000:1.

[9] Commonwealth Scientific and Industrial Research Organization.

[10] Typically, computer tests indicate the unconfined strength of a Synthetic Rock Mass of the order of 50-m to 100-m side length, to be a few percent of the unconfined strength of the laboratory specimen.

[11] A number of important subsurface engineering problems involve borehole access only. This often means difficulty in establishing reliable, realistic DFN's. In such cases there is no recourse, at least at the start of the project, other than to try to infer fracture networks from borehole observations, perhaps supplemented by local observations of structural geological features? The DFN for Northparkes was available and convenient to use in the example shown in Figure 5.

[12] Adapted from Fairhurst (1971).

[13] See also footnote 17 –Appendix 1.

[14] The URL at Bure was developed in order to determine the suitability of the Calllovo-Oxfordien Argillite formation for permanent storage of high–level nuclear waste.

[15] DAFSAM -Drilling Active Faults in South African Mines.

[16] http://www.icdp-online.org/front_content.php?idcat=460

ACKNOWLEDGEMENTS

Much of the material and concepts discussed in this paper is the result of work and discussions over many years with colleagues at Itasca Consulting Group, Inc. in Minneapolis and faculty in GeoEngineering at the University of Minnesota, especially in this instance, Professor Emmanuel Detournay. Particular help was received from Itasca colleagues Varun, Branko Damjanac, David Potyondy and Mark

Lorig, The influence of numerous stimulating discussions with Professor François Cornet of the Institut de Physique du Globe, Strasbourg, France are clearly evident in the paper. Professors Amos Nur and Mark Zoback, of Stanford University, USA and of Ingrain, Inc., Houston, USA assisted with valuable material, as acknowledged in the text. Dr. Rob Jeffrey and Andrew Bunger of CSIRO, Melbourne, the leaders in arranging HF2013, have provided valuable comments, assistance and understanding throughout. To all, I am very grateful. Such invaluable assistance notwithstanding, I accept full responsibility for the interpretations and views expressed in the paper.

REFERENCES

1. T. L Anderson, 2005Fracture Mechanics: Fundamentals and Applications. 3rd edition, CRC Press (0-84931-656-1
2. E. V Artyushkov, 1973Stresses in the Lithosphere Caused by Crustal Thickness
3. J Inhomogeneities, Geophy.Res. 7832November 10, 1973
4. A. P Bunger, X Zhang, and R. G Jeffrey, 2012Parameters Affecting the Interaction Among Closely Spaced Hydraulic Fractures" SPE Journal March 2012, 292306
5. M Calo, C Dorbath, F. H Cornet, and N Cuenot, 2011Large scale aseismic motion identified through 4D P-wave tomography; Geophys. J. Int. 18612951314
6. P. A Cundall, 2008An Approach to Rock Mass Modelling," in From Rock Mass to Rock Model-CD Workshop Presentations (15 September, 2008)- SHIRMS 2008 (Proc. 1st Southern Hemisphere International Rock Symposium, Perth, Western Australia, September 2008) Y. Potvin et al., Eds. Nedlands, Western Australia: Australian Centre for Geomechanics.
7. T. T Cladouhos, M Clyne, M Nichols, S Petty, W. L Osborn, and L Nofziger, 2011Newberry Volcano EGS Demonstration Stimulation Modeling" GRC Transactions, 35317322
8. F. H Cornet, 2012The relationship between seismic and aseismic motions induced by forced fluid injections." Hydrogeology Journal (2012) 20: 1463-1466
9. F. H Cornet, and T Röckel, 2012Vertical stress profiles and the significance of "stress decoupling". Tectonophysics 5812012193205

10. P. A Cundall, M. E Pierce, and D. Mas Ivars. (2008Quantifying the Size Effect of Rock Mass Strength" in SHIRMS 2008 (op.cit.) 2315

11. B Damjanac, and C Fairhurst, 2010Evidence for a Long-Term Strength Threshold in Crystalline Rock,|| Rock Mech. Rock Eng., 43, 513-531 (2010).

12. B Damjanac, C Detournay, P. A Cundall, and Varun, (2013Three-Dimensional Numerical Model of Hydraulic Fracturing in Fractured Rock Masses" Proc. HF 2013The International Conference for Effective and Sustainable Hydraulic Fracturing, Brisbane, May 20-22, 2013

13. B Damjanac, and C Fairhurst, Evidence for a Long-Term Strength Threshold in Crystalline Rock,|| Rock Mech. Rock Eng., 43, 513-531 (2010Duchane, D and D. Brown, (2002) "Hot Dry Rock (HDR) Geothermal Energy Research and Development at Fenton Hill, New Mexico" GHC (Geo-Heat Center) Bulletin, December. 2002 1319

14. C Fairhurst, and C Carranza-torres, 2002Closing the Circle- Some Comments on Design Procedures for Tunnel Supports in Rock," in Proceedings of the University of Minnesota 50th Annual Geotechnical Conference (February 2002), 2184J. F. Labuz and J. G. Bentler, Eds. Minneapolis: University of Minnesota, 2002. [available at www. itascacg.comgo to 'About'and Fairhurst Files]

15. C Fairhurst, 1971Fundamental Considerations Relating to the Strength of Rock. Colloquium on Rock Fracture, Ruhr University, Bochum, Germany, April 1971. (see http://www.itascacg.com/ about/ff.php)Revised and published in Report of the Workshop on Extreme Ground Motions at Yucca Mountain, August 23-25, 2004, U.S. Geological Survey, USGS Open-File Report 20061277T. C. Hanks et al., Eds. Reston, Virginia: USGS, 2006.

16. J Geertsma, and F De Klerk, 1969A Rapid Method of Predicting Width and Extent of Hydraulic Induced Fractures. J Pet Technol 211215711581SPE-2458-PA. http://dx.doi.org/10.2118/2458-PA

17. J. F Geyer, and S Nemat-nasser, 1982Experimental Investigation of Thermally induced Interacting Cracks in Brittle Solids Int. J. Solids and Structures 184349356

18. A. A Griffith, 1921The Phenomena of Rupture and Flow in Solids Phil. Trans. R. Soc. Lond. A 1921,, 221, 163-198 doi:rsta.1921.0006

19. A. A Griffith, 1924Theory of Rupture. Proc. First Int. Cong. Applied Mech (eds Bienzo and Burgers). 5563Delft: Technische Boekhandel and Drukkerij. 1924

20. G Grünthal, and R Wahlström, 2003An Mw-Based Earthquake Catalogue for Central, Northern and Northwestern Europe using a Hierarchy of Magnitude Conversions. J. Seismol. 7, 507-531 (Available

at http://seismohazard.gfzpotsdam.de/projects/en/eq_cat/menue_e"q_cat_e.html)

21. E Hoek, and Z. T Bieniawski, 1966Fracture Propagation Mechanism in Hard Rock," in Proceedings of the First Congress of the International Society of Rock Mechanics. Lisbon, September-October, 1243249J. G. Zeitlen, Ed. Lisbon: LNEC.
22. E Hoek, and E. T Brown, 1980Underground Excavations in Rock." Inst'n of Mining and Metallurgy (London) Revised 1982, 164
23. G. C Howard, and C. R Fast, 1970Hydraulic Fracturing" SPE Monograph 2. Henry L.Doherty Series 203 pp. SPE 30402
24. C. E Inglis, 1913Stresses in a Plate Due to the Presence of Cracks and Sharp Corners," Trans. Inst. Naval Arch., London, 55(1), 219141
25. R. G Jeffrey, et al2009Measuring Hydraulic Fracture Growth in Naturally Fractured Rock. SPE 124919; SPE Annual Technical Conference and Exhibition, New Orleans, Louisiana, USA, 47October 2009
26. J. F Knott, 1973Fundamentals of fracture mechanics, Wiley (0-47049-565-0
27. National Research Council2012Induced Seismicity Potential in Energy Technologies." Washington, DC: The National Academies Press, 2012. (300p.) View online at http://www.nap.edu/catalog.php?record_id=13355
28. J. M Nocquet, and E Calais, 2004Geodetic Measurements of Crustal Deformation in the Western Mediterranean and Europe " Pure Appl. Geophy., 161; 661668
29. R. P Nordgren, 1972Propagation of a Vertical Hydraulic Fracture. SPE J. 124306314SPE-3009-PA. http://dx.doi.org/10.2118/3009-PA.
30. R. B Peck, 1969Advantages and limitations of the observational method in applied soil mechanics. Geotechnique, 192171187
31. T. K Perkins, and L. R Kern, 1961Widths of Hydraulic Fractures. J Pet Technol 139937949SPE-89-PA. http://dx.doi.org/10.2118/89-PA.
32. W. S Pettitt, J. F Hazzard, B Damjanac, Y Han, M Pierce, T Katsaga, and P. A Cundall, Microseismic Imaging and Hydrofracture Numerical Simulations," in Proceedings, 21st Canadian Rock Mechanics Symposium (Alberta, Canada, May 5-9, 2012
33. M Pierce, 2011Discrete Fracture Network Simulation" DFN training session LOP (Large Open Pit). [ppt slides available on request. Itasca Consulting Group: www.itascacg.com]
34. A Riahi, and B Damjanac, 2013Numerical Study of Interaction between Hydraulic Fractures and Discrete Fracture Networks" Proc.

HF 2013The International Conference for Effective and Sustainable Hydraulic Fracturing, Brisbane, May 20-22, 2013

35. Royal Society and Royal Academy of Engineering (2012Junep. "Shale Gas Extraction in the UK: a review of hydraulic fracturing" Issued: June 2012, DES2597] View report online at: royalsociety.org/policy/projects/shale-gas-extraction and raeng.org.uk/shale
36. O Scotti, and F. H Cornet, 1994In-Situ Evidence for Fluid-Induced Asesismic Slip Events along Fault Zones. Int. J. Rock Mech Min.Sci. &Geomech. Abstr. 314347258Control in Mines (1965). South African Institute of Mining and Metallurgy, Johannesburg, 606p
37. A. M Starfield, and P. A Cundall, 1988Towards a Methodology for Rock Mechanics Modelling" Int. J. Rock Mech. Min. Sci.& Geomech. Abstr. 25 (3) 99106
38. J. F Tester, et al2006The Future of Geothermal Energy"-Impact of Enhanced Geothermal
39. Systems (EGS) on the United States in the 21st CenturyMIT Press.
40. J. D Wallis, J Devito, and E Diaz, 2012Digital Rock Physics- A New Approach to Shale Reservoir Evaluation" Oilfield Technology, March 2012 [http://www.ingrainrocks.com/articles/a-new-approach-to-shale-reservoir-evaluation/]
41. X Zhang, and R. G Jeffrey, 2008Re-initiation or termination of fluid-driven fractures at frictional bedding interfaces" JGR, 113BO 8416, doi:10.1029/2007JB005327,

Chapter 6

LIQUID ENCAPSULATION TECHNOLOGY FOR MICROELECTROMECHANICAL SYSTEMS

Norihisa Miki[1]

[1]Department of Mechanical Engineering, Keio University, Hiyoshi, Kohoku-ku, Yokohama, Kanagawa, Japan

INTRODUCTION

Microelectromechanical systems (MEMS) have been extensively studied for over three decades, which has resulted in the prevalence of quite a few commercially available MEMS products in our daily lives, although they are too small to see. In the very beginning of the MEMS success story, people recognized the importance of packaging [1]. MEMS contain mechanical parts, and given their small sizes, they are severely affected by surrounding molecules. Therefore, MEMS are packaged under vacuum, at low pressure, or at least free from water molecules. Water molecules can bridge two separated parts and bring them into contact by the meniscus force, which may lead to permanent adhesion of the parts, known as stiction. This phenomenon must be averted, not only in the packaging, but also in the fabrication of parts. It is not an overstatement to say that researchers go to great lengths to keep their devices dry.

On the other hand, as MEMS technologies advance, a wide variety of applications are expected, some of which the MEMS must handle liquids. For example, drug delivery systems (DDS) that administer medicine to diseased parts at designated times can employ MEMS that are sufficiently small to be implanted and are capable of controlling discharge of the medicine [2-5]. In this application, the MEMS must contain medicine, which is in liquid form in many cases. In addition, MEMS can be used as a portable power source, referred to as power MEMS. Micro gas turbines and certain fuel cells require liquid fuel to generate chemical reactions [6-9]. Micro total analysis systems, or microTAS, are used to manipulate minute aqueous analytes and/or control microfluids to handle samples, such as cells and bacteria, for biochemical analysis [10-14].

Such useful characteristics of liquids are available to expand the design space for innovative MEMS devices. Functional liquids, such as magnetorheological fluid and electroconjugate liquids, can be used as micro pumps and actuators [15-16]. Other useful characteristics of liquids include deformability, incompressibility, and high dielectric constant. Hydraulic amplification can be achieved by exploiting the deformability and incompressibility of liquids [17-22]. In addition, highly dielectric liquids can enhance sensor sensitivity while maintaining flexibility [23].

While some applications allow such MEMS devices to bring the liquid from outside, encapsulation of liquid inside MEMS devices is mandatory in other applications. Liquid encapsulation technology can be used to manufacture innovative MEMS devices, such as completely spherical microlenses and hydraulic amplification mechanisms. Various liquid encapsulation technologies have been proposed to achieve these promising applications. The liquid species to be encapsulated and the application must be taken into consideration for the selection of appropriate encapsulation processes. This chapter reviews state-of-the-art liquid encapsulation technologies and their application to the manufacture of innovative MEMS devices that exploit the useful characteristics of the encapsulated liquids.

FILL AND SEAL TECHNIQUE

The most straight-forward technique to encapsulate liquids is to dispense liquid into a reservoir and then seal it with another substrate, as shown in Figure 1. The reservoirs can be easily manufactured using conventional MEMS fabrication technologies. Currently, commercially available dispensers are capable of dispensing a minute amount of liquid, as small as several nanoliters. Sealing, or bonding the substrates, is the most critical process.

In the MEMS field, bonding technologies have been widely explored for packaging and manufacturing three-dimensional structures [24-27]. The direct bonding of silicon wafers, anodic bonding of glass substrates, and thermocompression bonding using a metal thin film as an adhesive layer are examples of frequently used technologies. These technologies achieve strong bonding of substrates via covalent bonds; however, these processes have drawbacks when applied to the sealing of a liquid filled reservoirs if they require high temperatures. High temperature processes in the order of several hundreds degrees Celsius may change the properties of the liquid to be encapsulated. For example, in drug delivery systems, it is necessary to maintain the medicinal effect of the encapsulated drug. Fuels for power MEMS applications should not be burnt before the device is completed. In addition, some liquids are volatile, which precludes not only high-temperature, but also vacuum processes.

Therefore, in this fill and seal approach, adhesive bonding shown in Figure 1 is the most appropriate technique. The adhesives employed include epoxy, UV curable resin including photoresist, and benzocyclobutene (BCB). Such adhesives solidify after either mixing with curing agents, exposure to UV irradiation, or thermal treatment at low temperatures. Either one-part or two-part epoxy resins can be used, because they do not require high temperature to promote solidification. The chemical reaction progresses with time, even at room temperature, and after solidification, the epoxy achieves strong bonds and is resistant to many chemicals.

While epoxy is a common adhesive, it is not compatible with conventional MEMS fabrication technology. To achieve reliable and reproducible bonding, the adhesives are preferably spin-coated, which allows the thickness to be controlled according to the spinning speed. In this regard, photoresist is a good candidate. Photoresists are compatible with MEMS fabrication technologies and can be spin-coated, and more importantly, knowledge of their use is well developed. Photoresist is coated on a substrate and then brought into contact with the pairing substrate either before or after the curing processes. A typical curing temperature is around 100 °C. When the contact is performed after curing, the bonding is achieved by a hot melt process at higher temperatures, although lower than 200 °C. Photoresist can be patterned using conventional photolithography to determine the bonding areas. The major drawback of using photoresists as adhesives is the weakness of the bond strength, i.e., they are not designed to function as adhesives. Adhesion between the photoresist and the substrates, as well as the mechanical strength of the photoresist, is designed to be sufficiently strong to survive photolithography processes. Therefore, the bonding may fail due to external forces, either at the interface or within the bulk.

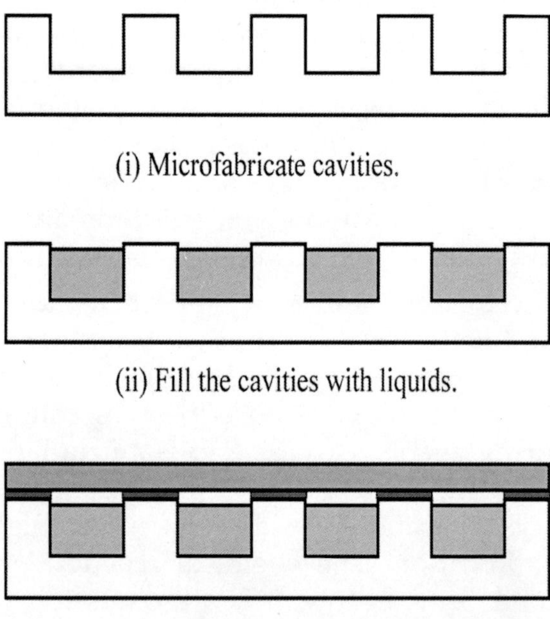

FIGURE 1. Fill and seal approach.

BCB is a promising polymer adhesive that is photo-patternable and compatible with conventional photolithography processes. It can be spin-coated to thicknesses of 5-15 μm at spinning speeds of 1000-6000 rpm [28]. BCB has good chemical resistance, and the most significant advantage of this material is that it does not release any gases during the cure, which does not create pores in the material or contaminate the encapsulated liquid. BCB can be used to bond two substrates by thermocompression bonding. Compression at 230 °C has been attempted, which may limit the species of liquid to be encapsulated. However, BCB has been applied to seal sodium hypochlorite aqueous solution (NaOCl) for galvanic cells [28]. The paper discusses the BCB thickness and the bond quality determined by the geometry of the bonding areas. UV curable resins do not require heat treatment, but only UV irradiation. If the MEMS devices are not UV sensitive and one substrate is transparent to UV light, then UV curable resin offers a strong bond after solidification with UV irradiation. Such bonding can even be conducted in liquids [21,22,29] and we have termed this the bonding-in-liquid technique (BiLT).

We have introduced sealing processes that employ polymer adhesives. However, the gas permeable nature of polymers may cause problems of contamination and vaporization of volatile liquids. For example, polydimethyl siloxane (PDMS), which is one of the most frequently used polymers in the fields of MEMS and microTAS, is permeable to gas. However, this permeability can be modified by the addition of different materials [30] or coating with airtight films [31]. Typical polymers are several orders of magnitude more permeable to gas than metals and ceramics [27]. Therefore, sealing with gold stud bumps has been proposed [32], where reservoirs are filled up with the liquids via microchannels and the inlets and outlets of the channels are then plugged with wire-bonding gold. Firstly, a gold ball is formed at the edge of the gold wire by electrical discharge. The ball is then pressed to the opening of the channel using ultrasound. The wire is then cut and the sealing is completed. Helium leak tests were conducted and hermetic sealing was verified using this technique when the hole diameters were less than 40 μm.

The inevitable drawback of the fill and seal approach is the filling rate; it is quite difficult to completely fill a reservoir with a liquid. This is acceptable for some applications, such as drug delivery and fuel supply for power MEMS devices. However, the performance of hydraulic displacement amplification mechanisms (HDAM) is deteriorated by the interfusion of compressible air. When liquids are used as components of sensors, contamination of gas or other liquids will lead to a loss of sensitivity. Therefore, liquid encapsulation techniques that enable complete filling of liquids are mandatory. The author's group developed BiLT, which is a fill and seal approach that enable complete filling [21,22,29].

BONDING-IN-LIQUID TECHNIQUE (BILT)

HDAMs require complete filling of the reservoir with an incompressible liquid, because gas is much more compressible than liquid. Figure 2 shows a package chamber that has openings at the top and bottom, where incompressible liquid is encapsulated with flexible polymer membranes. The top opening, which is determined by a metal plate, is smaller than the bottom opening; therefore, a small displacement applied to the bottom membrane is amplified at the top, according to the ratio of the openings. The application of HDAM is discussed in section 5.2. The key points in the fabrication of HDAMs are no interfusion of air bubbles and sealing with flexible polymer membranes. Complete filling can be achieved by the direct deposition of a thin film, which is detailed in the following section; however, this technique does not allow the use of flexible membranes. We have developed BiLT [29], which can be employed to overcome this problem.

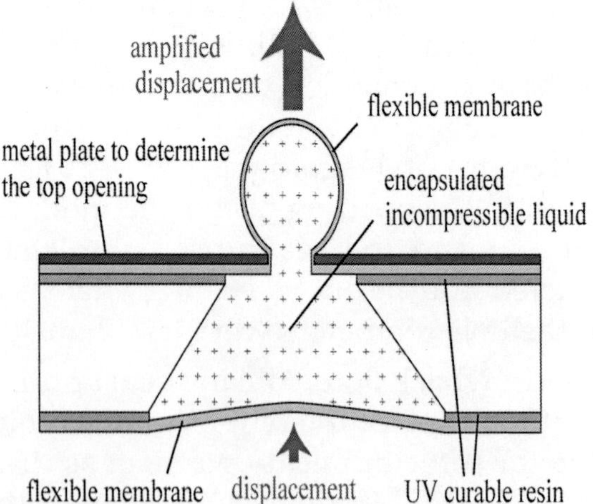

FIGURE 2. Schematic cross-sectional view of HDAM.

Rather than package the MEMS devices vacuum, we considered that if the encapsulating process was conducted in liquid, then the reservoir could be filled without the interfusion of air bubbles. However, one concern was how to successfully bond the membrane to the package chamber in a liquid environment. Therefore, it was decided to use a UV-curable resin (3164 Three Bond, Three Bond Co., Ltd.) that is solidified after UV irradiation, even in a liquid environment. This membrane achieves a tensile strength of 0.85 MPa when cured and the thickness of the resin can be controlled according to the spin-coating speed.

Figure 3 depicts the procedures employed in BiLT. Firstly, a UV resin is coated onto the bonding surface; however, it should be noted that the surface has many cavities for liquid encapsulation. Therefore, the UV resin is spin-coated onto a thick PDMS membrane and then transferred onto the bonding surface by soft contacting the PDMS membrane (Figure 3(a,d)). A sufficient amount of resin needs to be applied to the bonding surface to achieve a good bond, while excess resin may fall into and occupy the cavity during the bonding process. UV resin thicknesses of 80, 120, 160, and 200 μm were tested on PDMS membranes, which correspond to spin-coating speeds of 4000, 3000, 2000, and 1000 rpm, respectively. When silicon was used as the bonding substrate, the transferred thicknesses were 7.9, 8.1, 17, and 27 μm. In case of UV resin thicknesses of 17 and 27 μm, the excess resin flowed into the cavity.

Handling of a flexible thin membrane is not a trivial process. The membrane must be kept flat throughout the bonding process. Therefore, the membrane was spin-coated and cured on a glass substrate. The thickness of the PDMS membrane can be controlled according to the spin-coating speed. During the bonding process, the PDMS membrane must be peeled off the glass substrate. Therefore, the glass surface is coated in advance with a hydrophobic film (CYTOP M, H, Asahi Glass Corporation) to facilitate exfoliation.

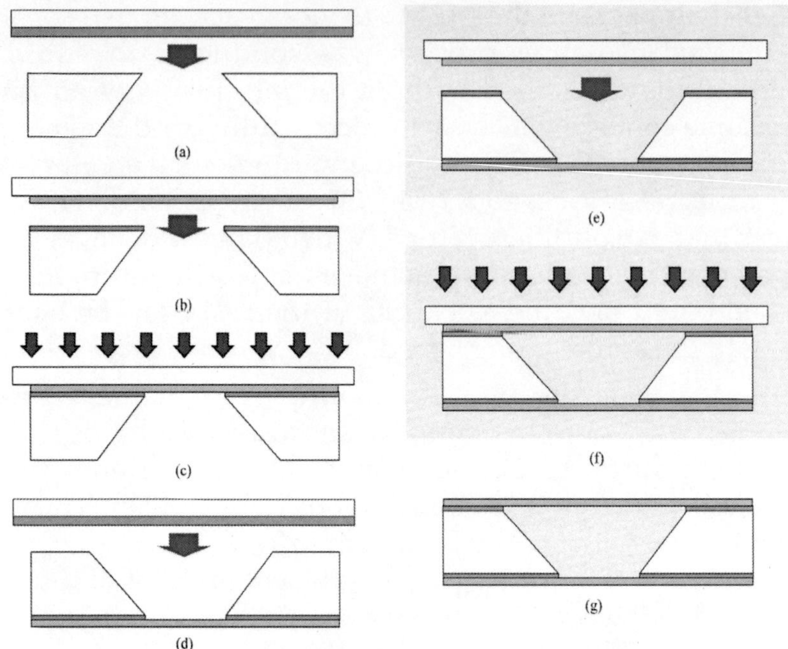

FIGURE 3. BiLT process. (a) UV curable resin is transferred onto the bonding surface. (b) A flexible membrane coated onto another substrate in advance is brought into contact with the bonding surface. The membrane is coated onto a hydrophobic layer to facilitate peeling of the membrane from the substrate. (c) UV light is irradiated to cure the UV resin. (d) A second UV curable resin is transferred onto the bonding surface. (e) A flexible membrane is brought into contact with the bonding surface in a liquid environment. (f) UV light is irradiated to cure the resin. (g) Liquid encapsulation without the interfusion of air bubbles or deformation of the membrane is achieved.

The substrate with cavities and the flexible membrane on the glass substrate are brought into contact in a liquid environment (Figure 3(e)). UV light is then irradiated onto the bonding surface through the glass substrate and flexible membrane to cure the UV curable resin (Figure 3(f)). Note that the substrate and membrane must be UV-transparent for this process. Figure 4 shows micrographs that confirm liquid (red-dyed deionized (DI) water) encapsulation was completed without the interfusion of air bubbles. Excess resin flowed into the cavities for UV resin thicknesses of 17 and 27 μm. No DI water was observed at the bonding interface. Encapsulation of glycerin was also attempted. Glycerin is non-volatile, so that the volume of the encapsulated glycerin did not change over a period of weeks even when encapsulated with a gas permeable PDMS membrane at ambient pressure and room temperature.

In HDAM, encapsulated liquids are sealed with flexible membranes at both the top and bottom sides of the package chamber. When the encapsulation/bonding process is conducted in air and not in liquid, the difference in the density of the air and liquid result in bowing of the membrane. Note that the membranes must be kept flat during the bonding processes of BiLT.

The bond strengths were investigated by conducting 180° peel tests on PDMS membranes and silicon substrates bonded using BiLT in DI water, glycerin, phosphate buffer solution (PBS), isopropyl alcohol (IPA), and acetone, and also in air as a reference. The silicon substrate used in the experiments did not contain bonding cavities. The bond strengths of the samples were measured as a function of time (1, 6, 24, 72, and 168 h) using a dynamic mechanical analyzer (RSAIII, TA Instruments). The test procedure involved one edge of the PDMS membrane being manually peeled from the silicon substrate and the unbonded area of the silicon substrate being clamped. The peeled PDMS membrane was then pulled in the direction parallel to the bonding interface at a speed of 3 mm/min until it peeled off, and the shear stress required to peel the PDMS membrane from the silicon substrate was measured. The results are shown in Figure 5. The bonding resin was dissolved in both IPA and acetone solution, and thus bonding was unsuccessful when conducted in these solutions, while the bonding strengths of the other samples were comparable. The bond strengths increased with time, most likely due to continuing chemical reaction of the UV-curable resin over time. The bonding strengths after 1 week were more than 4 times greater than those obtained after 6 h. Peel tests conducted within 72 h of bonding revealed failure of the resin, while failure occurred at the interface between the resin and PDMS membrane when measured after 1 week. This indicates that failure occurred within the resin until the resin was sufficiently cured and this is why the bonding strengths in air, DI water and PBS were similar; 1 week after bonding, the bonding strengths achieved by bonding in air and using BiLT were comparable.

The developed BiLT enables complete liquid filling with various membranes. Many species of liquids can be encapsulated using

BiLT, unless the liquids dissolve the UV resin. This feature is crucial in manufacturing HDAM and sensors, which will be introduced in section 5. Complete filling can be achieved by direct deposition of a thin film, as introduced in section 4; however, this process can only be used to encapsulate non-volatile liquids, and the type of sealing membrane is also limited. The major drawback of this technology is that the substrate must be UV-transparent and the device should not contain UV-sensitive materials. For example, dye-sensitized photovoltaic cells, which are employed as transparent solar cells and optical sensors, require the encapsulation of electrolytes. However, BiLT cannot be used for encapsulation, because the cells have dyes that degrade after being exposed to UV.

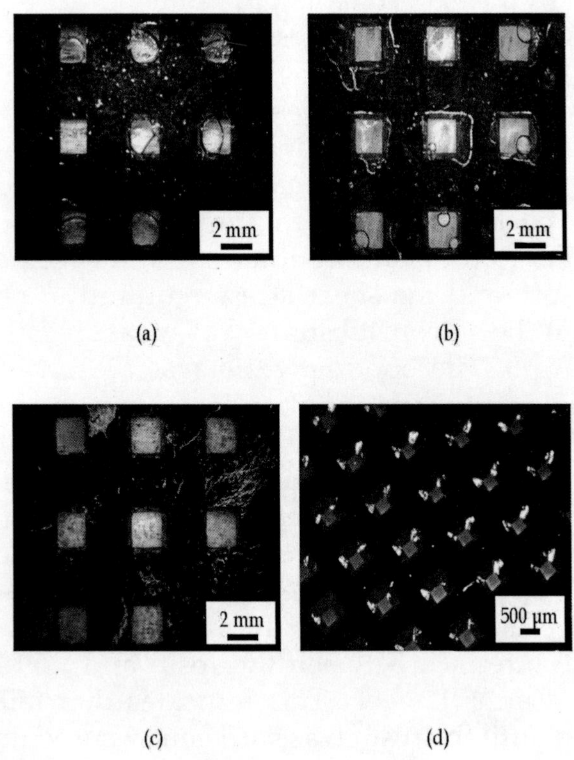

FIGURE 4. Bonding results for resin thicknesses of (a) 27, (b) 17 and (c) 8.1 μm. Red-dyed water was encapsulated into the cavities. The transparent parts in the cavities are excess resin. When a certain amount of resin was used, excess UV resin or air was found in the cavities, as shown in (d).

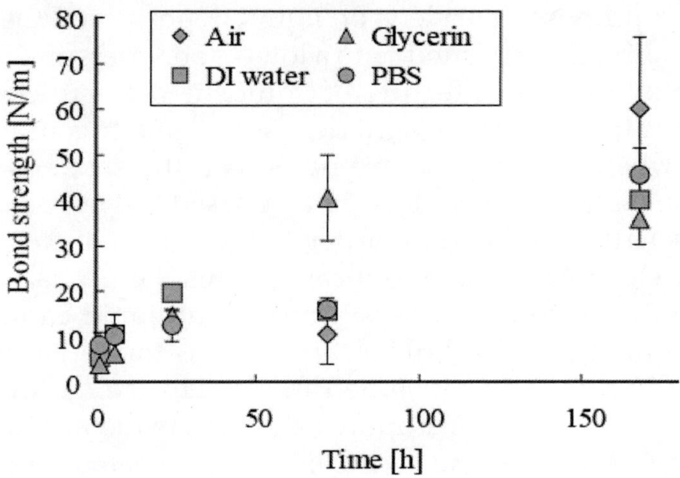

FIGURE 5. Bond strength after BiLT in different media as a function of time.

LIQUID ENCAPSULATION BY DIRECT DEPOSITION OF A THIN FILM

Thin film deposition onto a solid is typically conducted under vacuum. Some liquids, such as silicone oil and ionic liquids, have extremely low vapor pressure and do not evaporate under vacuum. A thin film of metal or polymer can be directly deposited onto such low-vapor-pressure liquids. A thin silver film was deposited onto an ionic liquid to manufacture a mirror for a space telescope [33].

Parylene, or poly(para-xylylene), is widely used in fields of MEMS and microTAS due to favorable characteristics, such as transparency, mechanical strength (3.2 GPa), biocompatibility, gas sealing efficacy, and has the ability to be conformally coated using chemical vapor deposition [34-36]. Parylene has been used to form a microspring with a low spring constant [37] and as a substrate and/ or a protective layer for the manufacture of microelectrodes [38]. In addition, the surfaces of PDMS microchannels have been coated with parylene to protect against protein adsorption [39].

The typical pressure for parylene deposition is several pascals. Therefore, liquids with vapor pressures less than this can remain in the liquid phase during the parylene deposition process. Binh-

Khiem et al. proposed parylene on liquid deposition (POLD), where parylene is directly deposited onto a low-vapor-pressure liquid, such as silicone oil [34-36]. A feature of liquids is that sufficiently small droplets can have perfectly spherical shape due to surface tension; however, when the droplets are not small, the shape is deformed by gravity. Spherical droplets can be used as lenses. The focal length of a lens is in the same order as the lens diameter; therefore, small lenses enable compact optical systems. POLD can be used to form spherical microlens arrays; transparent liquid can be directly deposited onto silicone oil droplets. A film is formed at the droplet surface, so that no air is included in the lens, i.e., a filling rate of 100% is achieved. Parylene is more flexible than metals; therefore, by integrating electrodes beneath the liquid lens coated with parylene, the lens can be deformed to vary the focal length according to the voltage applied to the electrodes [34].

The direct deposition on liquid approach enables perfect liquid encapsulation with good reproducibility. The useful characteristics of parylene or specific metals can be exploited; however, the disadvantage is that the liquids to be encapsulated must be non-volatile and the type of sealing material is limited. For example, this approach cannot be applied to HDAM, because it is preferable to seal the liquid with flexible membranes.

APPLICATIONS OF LIQUID ENCAPSULATION TECHNOLOGY

Drug Delivery

In DDS, the release of medicine is designed so that the drug efficacy is high. MEMS-based DDS are expected to convey medicine to the vicinity of diseased sites by exploiting the small MEMS size and drug release occurs when appropriate. Santini et al. proposed the controlled release of drugs from cavities that were sealed with a thin gold layer [2]. The thin gold layer is used as an anode and an electrochemical reaction occurs, with subsequent dissolution of the gold film and release of the encapsulated medicine. Drug delivery is thus controlled by electrical control of the reaction.

In this application, the fill and seal approach is employed to maintain the quality of the medicine as a priority. The package chambers are prepared with a thin gold layer as the bottom surface. Drugs are dispensed into the chambers and then sealed with a water-proof membrane using epoxy adhesive. Details of this process can be found in the literature [2].

HYDRAULIC AMPLIFICATION

Some MEMS applications require large displacement of several tens of micrometers. High-flow-rate microvalves are one such application. To satisfy the requirements, hydraulic amplification has been reported that utilizes incompressible fluid in a microchamber with an input surface that is larger than the output surface [17-20]. Another example is a tactile display [21,22]; Figure 6 shows an illustration of an array of MEMS actuators that mechanically deform the fingertip and stimulate tactile receptors. These receptors typically require skin deformation of several tens of micrometers, which could be facilitated by hydraulic amplification. Tactile displays are developed to offer a new approach to human-machine interface for virtual reality applications and interactive devices such as pointing devices or game controllers, and also to support visually impaired persons. In the microvalve applications, a working fluid can be used as an incompressible fluid for hydraulic amplification. However, tactile displays cannot afford to have an external fluidic system to drive a working fluid; therefore, complete encapsulation of the liquid is necessary. In addition, the sealing membranes must be flexible. HDAMs have been successfully developed by encapsulating incompressible and non-volatile glycerin in microchambers with flexible and largely deformable PDMS membranes (mixture of DC 3145 CLEAR and RTV thinner, Dow Corning Toray Inc.) via UV curable resin using BiLT. Figure 2 shows a schematic of a HDAM [21,22].

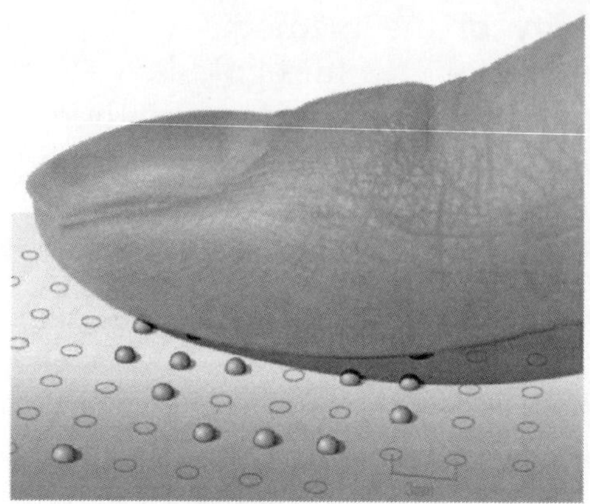

FIGURE 6. Tactile display that employs an array of MEMS actuators.

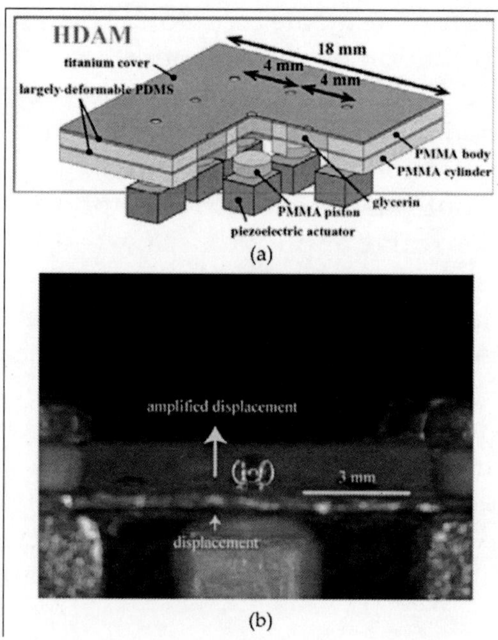

FIGURE 7. Large displacement MEMS actuator array that consists of HDAM with glycerin-filled cavities prepared using BiLT, and piezoelectric actuators. (a) Schematic image of HDAM, and (b) micrograph showing the large deformation of a PDMS membrane.

The HDAM shown in Figure 7(b) was combined with piezoelectric actuators and applied to develop large-displacement MEMS-actuators, with a particular aim to application in MEMS-based tactile displays [21,22,40]. When applied to a vibrational Braille code display, it was experimentally verified that the large-displacement HDAM could display Braille codes more efficiently than a static display. This is because both fast and slow adaptive tactile receptors could be used to detect the displayed patterns when individual cells were vibrated at several tens of hertz [40]. When the actuation of the large displacement MEMS actuators was controlled both spatially and temporary, different surface textures, such as rough and smooth, could be displayed.

FLEXIBLE CAPACITIVE SENSOR

Highly sensitive pressure sensors are expected to be applied in humanoid robots and medical instruments to detect tactile sensation, which would enable safe physical interaction with the environment, including human contact. MEMS-based capacitive sensors that have simple structures composed of electrodes and a dielectric component have been widely studied, due to good compatibility with MEMS fabrication technologies. Capacitive sensors require not only high sensitivity, but also flexibility to detect the pressure applied to curved surfaces. Silicon-based MEMS capacitive sensors have been developed; however, silicon is brittle, which makes it difficult for the sensors to conform to a curved surface. Therefore, polymer-based flexible sensors have been proposed and demonstrated. Polymer-based flexible sensors are typically used to maintain flexibility with air as the dielectric; however, air has a relatively low dielectric constant. A solid dielectric may enhance the sensitivity, but impairs the flexibility of the sensor. Therefore, a polymer-based capacitive sensor that uses a dielectric liquid has been proposed, as depicted in Figure 8 [23]. DI water and glycerin have high relative dielectric constants of approximately 80.4 and 47; therefore, the proposed sensor with such liquids can have high sensitivity while maintaining flexibility. The capacitance of the electrodes increases when pressure is applied to the device. PDMS is used as a structural material in this device. An escape reservoir is designed to allow an incompressible liquid, such as DI water and glycerin, to move from the cavity between the electrodes when pressure is applied to the sensor, which allows the flexible sensors to deform and vary the capacitance. The proposed microsensor has been fabricated, and both high sensitivity and flexibility have been experimentally demonstrated.

DYE-SENSITIZED PHOTOVOLTAIC CELL

Dye-sensitized photovoltaic cells are currently attracting widespread scientific and technological interest as a high efficiency, low-cost, and transparent alternative to inorganic solar cells. Figure 9shows a schematic illustration of the structure and operation principle of the dye-sensitized photovoltaic device. The cell consists of two electrodes and an encapsulated liquid electrolyte that contains iodide and triiodide ions. The cathode is a highly porous nanocrystalline semi-conductive titanium dioxide (TiO_2) layer, in many cases consisting of TiO_2 nanoparticles, deposited on a transparent electrically conductive glass. TiO_2 absorbs only UV light; therefore, dye is adsorbed onto the TiO_2 layer to utilize the light with a wider range of wavelength. The counter electrode (anode) is a transparent electrically conductive glass with a platinum catalyst. The device is transparent and is colored according to the dye employed.

When light passes through the electrically conductive transparent glass electrode, the dye molecules are excited and transfer an electron to the semiconducting TiO_2 layer via electron injection. The electron is then transported through the TiO_2 layer and collected by the conductive layer on the glass. The mediator (I-/I3-) undergoes oxidation and regeneration in the electrolyte. Electrons lost by the dye molecules to the TiO_2 layer are replaced by electrons from the iodide and triiodide ions in the electrolyte, thereby generating iodine or triiodide, which in turn obtains electrons at the counter electrode, culminating in a current flow through the external electrical load. This is the mechanism for the conversion of light energy received by the device to electricity [41]. This device has an interesting feature in that it reacts strongly to light that enters through the TiO_2 layer.

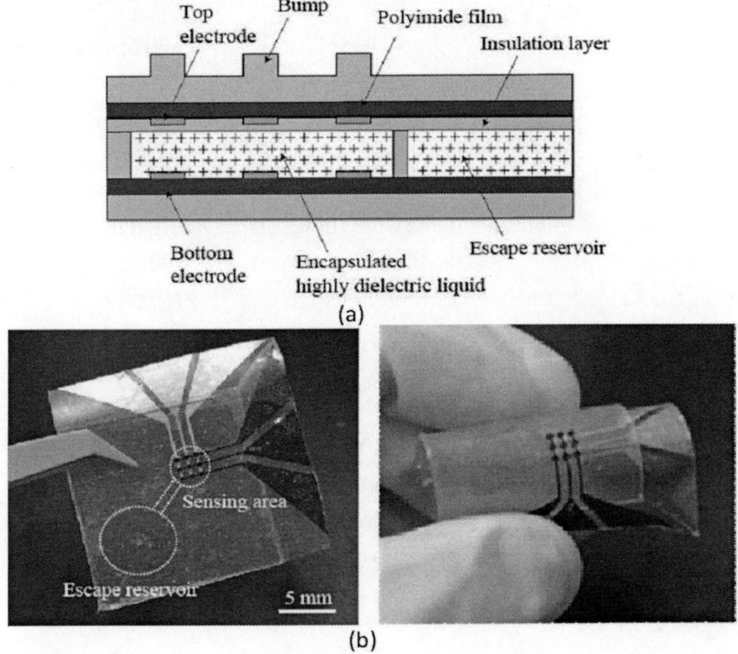

FIGURE 8. a) Cross-sectional view of a capacitive sensor with encapsulated liquid dielectric to enhance the sensitivity while maintaining flexibility. (b) Fabricated capacitive sensor.

The dye-sensitized photovoltaic cell has been conventionally studied as a solar cell, where miniaturization was not considered. However, when the cells are microfabricated and arrayed, they can be used as a transparent optical sensor. Shigeoka et al. proposed to microfabricate a transparent optical sensor on eyeglasses, which could detect the pupil position by detecting reflection from the eye, as shown in Figure 10 [42]. The light reflected from the pupil is considered to be smaller than that from the white. The sensor reacts strongly to light from the TiO_2 electrode side, i.e., when the TiO_2 layer electrode is faced towards the eyes, it detects only the light reflected from the pupil and white of the eye, without being affected by the light incident on the device from the environment.

Figure 11 shows a schematic of the processes used for fabrication of this device. The most critical part is encapsulation of the electrolyte. The conductive layer (ITO) is firstly patterned on the glass substrate using photolithography. TiO_2 nanoparticles are patterned on the cathode using a lift off process. The device is subsequently annealed in air at 450 °C for 60 min and then dipped in a ruthenium-containing dye solution for 60 min to ensure the dye is adsorbed onto the TiO_2nanoparticles. The two glass substrates are bonded via a hot melt film and application of 600-800 kPa at 100 °C. Lastly, the liquid electrolyte is flowed from the inlet hole into a channel between the two electrodes, and then the inlet and outlet holes are covered by end seals. The dyes used are UV sensitive; therefore, BiLT was not applicable to this liquid encapsulation process, and the fill and seal approach was used instead. However, the filling rate of the electrolyte was quite high and no interfusion of air between the electrodes was observed. An array of the dye-sensitized photovoltaic devices successfully detected the pupil position. The line-of-sight (LOS) was successfully deduced [42, 43] from the obtained pupil position and the front image of the subject, acquired using a CCD camera on the eyeglasses.

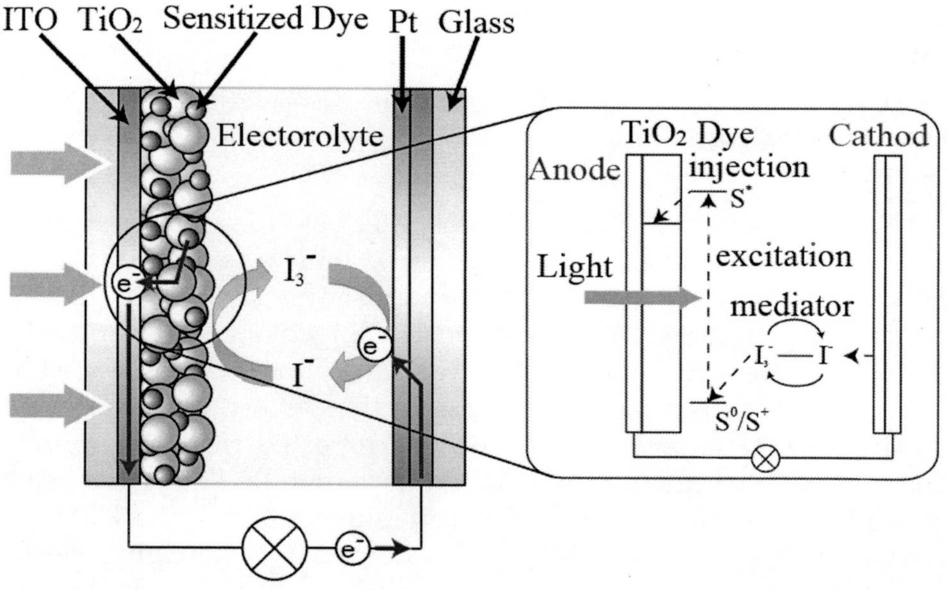

FIGURE 9. Structure and operation principle of the dye-sensitized photovoltaic device.

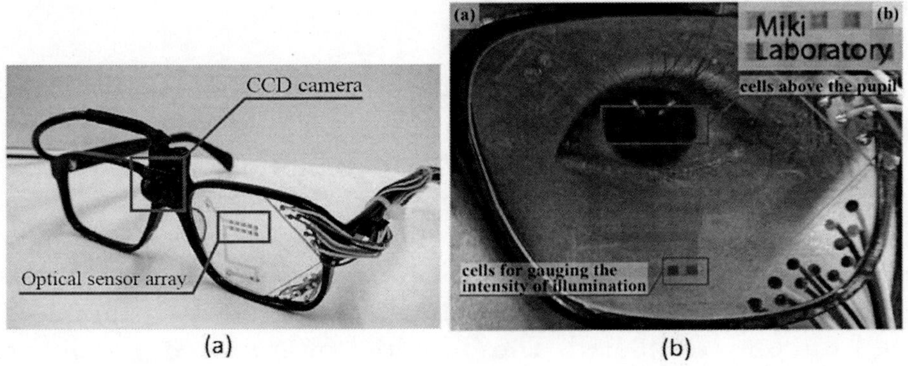

FIGURE 10. a) Array of dye sensitized photovoltaic cells patterned onto eyeglasses to detect the pupil position. The electrolyte was encapsulated between the electrodes. (b) Photograph of the sensor when worn by a subject.

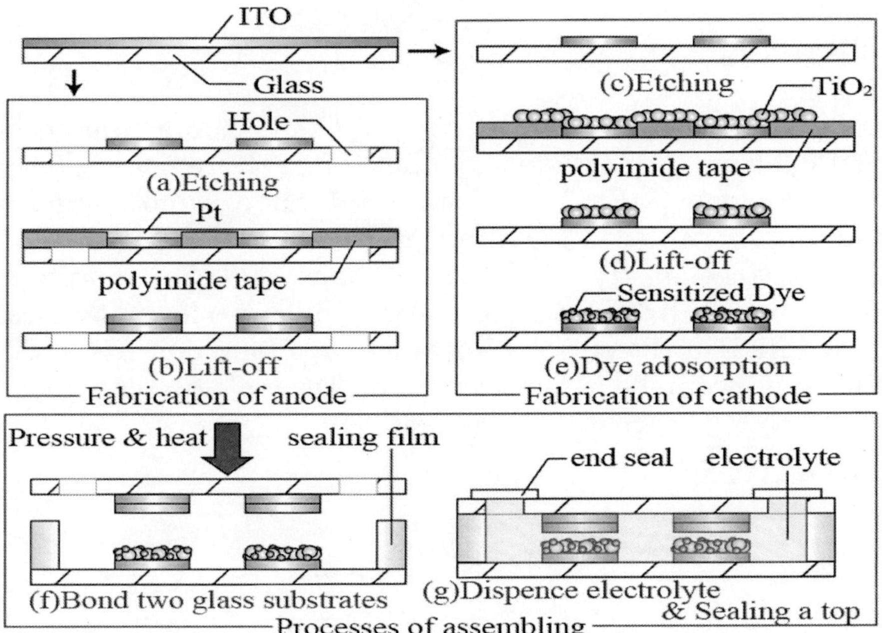

FIGURE 11. Fabrication process to produce an array of dye-sensitized photovoltaic devices. The fill and seal approach was employed to encapsulate the electrolyte (g).

CONCLUSION

This chapter has reviewed liquid encapsulation technologies and their applications to the manufacture of innovative MEMS devices that exploit the useful characteristics of liquid; liquid is deformable and liquid droplets form a perfectly spherical shape by surface tension. Other liquids have high relative dielectric constants. Liquids can be used as drugs for DDS and fuels for power MEMS. Appropriate liquid encapsulation technologies must be selected according to the liquid to be encapsulated. The fill and seal approach, bonding-in-liquid technique, and direct deposition of a thin film were discussed in this chapter, all of which have both advantages and disadvantages.

The use of liquid in MEMS packaging is quite a new technology. The author is convinced that more and more liquid encapsulation technologies will be developed and contribute to the further development of innovative liquid-encapsulating MEMS devices.

Acknowledgements

This work was supported by a Grant-in-Aid for Young Scientists (B) (21760202) from the Ministry of Education, Culture, Sports, Science and Technology (MEXT) of Japan, the Strategic Information and Communications R&D Promotion Programme (SCOPE) (092103005) of the Japan Ministry of Internal Affairs and Communications (MIC), and the Information Environment and Humans research area of PRESTO (Precursory Research for Embryonic Science and Technology) from the Japan Science and Technology Agency (JST).

REFERENCES

1. S. D Senturia, Microsystem Design. Kluwer Academic Publishers; 2001
2. J. T Santini, M. J Cima, R Langer, A controlled-release microchip. Nature 1999397335338
3. J. T Santini, A. C Richards, R Scheidt, M. J Cima, R Langer, Microchips as controlled drug-delivery devices. Angewandte Chemie International Edition 20003923962407
4. Y Li, R. S Shawgo, B Tyler, P. T Henderson, J. S Vogel, A Roesnberg,

P. B Storm, R Langer, H Brem, M. J Cima, In vivo release from a drug delivery MEMS device. Journal of Controlled Release 2004100211219

5. Richards Grayson ACShawgo RS, Li Y, Cima MJ. Electronic MEMS for triggered delivery. Advanced Drug Delivery Reviews 200456173184
6. A. H Epstein, S. D Senturia, Macro power from micro machinery. Science 1997
7. N Miki, C. J Teo, L. C Ho, X Zhang, Enhancement of rotordynamic performance of high-speed micro-rotors for power MEMS applications by precision deep reactive ion etching. Sensors and Actuators A: Physical 2003104263267
8. J. D Holladay, E. O Jones, M Phelps, J. L Hu, Microfuel processor for use in a miniature power supply. Journal of Power Sources 2002
9. L. R Arana, S. B Schaevitz, A. J Franz, M. A Schmidt, K. F Jensen, Journal of Microelectromechanical Systems 2003125600612
10. A Manz, N Graver, H. M Widmer, Miniaturized total chemical-analysis systems- a novel concept for chemical sensing. Sensors and Actuators B: Chemical; 1244248
11. S Shoji, Fluids for sensor systems. Topics in Current Chemistry 1998194163168
12. K Huikko, R Kostiainen, T Kotiaho, Introduction to micro-analytical systems: bioanalytical and pharmaceutical applications. European journal of pharmaceutical sciences 200320149171
13. H Ota, T Kodama, N Miki, Rapid formation of size-controlled three-dimensional hetero-cell aggregates using micro-rotation flow for spheroid study. Biomicrofluidics 2011
14. Y Gu, N Miki, Multilayered microfilter using a nanoporous PES membrane and applicable as the dialyzer of a wearable artificial kidney. Journal of Micromechanics and Microengineering 2009
15. S Yokota, K Kawamura, K Takemura, K Edamura, High-integration micromotor using electro-conjugate fluid (ECF). Journal of Robotics and Mechatronics 2005172142148
16. A Yamaguchi, K Takemura, S Yokota, K Edamura, A robot hand using electro-conjugate fluid. Sensors and Actuators A: Physical 2011170139146
17. D. C Roberts, L Hanqing, J. L Steyn, O Yaglioglu, S. M Spearing, M. A Schmidt, N. W Hagood, A piezoelectric microvalve for compact high-frequency, high-differential pressure hydraulic micropumping systems. Journal of Microelectromechanical Systems 20031218192
18. H Kim, K Najafi, Electrostatic hydraulic three-way gas microvalve for high-pressure applications. Proceedings of the 12th International

Conference on Miniaturized Systems for Chemistry and Life Sciences, MicroTAS 20081216October 2008San Diego, USA.

19. H Kim, K Najafi, An electrically-driven large-deflection high-force, micro piston hydraulic actuator array for large-scale microfluidic systems. Proceedings of 22th IEEE International Conference on Micro Electro Mechanical Systems, MEMS 20092529January 2009Sorrento, Italy.
20. X Wu, S. H Kim, C. H Ji, M. G Allen, A piezoelectrically driven high flow rate axial polymer microvalve with solid hydraulic amplification. Proceedings of 21st IEEE International Conference on Micro Electro Mechanical Systems, MEMS 20081317January 2008Tuscon, USA.
21. X Arouette, Y Matsumoto, T Ninomiya, Y Okayama, N Miki, Dynamic characteristics of a hydraulic amplification mechanism for large displacement actuators systems. Sensors 20101029462956
22. T Ninomiya, Y Okayama, Y Matsumoto, X Arouette, K Osawa, N Miki, MEMS-based hydraulic displacement amplification mechanism with completely encapsulated liquid. Sensors and Actuators A: Physical; 166277282
23. Y Hotta, Y Zhang, N Miki, A flexible capacitive sensor with encapsulated liquids as dielectric. Micromachines 20123137149
24. Q. Y Tong, U Gosele, Semiconductor wafer bonding science and technology: John Wiley & Sons, Inc.; 1999
25. N Miki, Wafer bonding techniques for MEMS. Sensor Letters 200534263273
26. N Miki, S. M Spearing, Effect of nanoscale surface roughness on the bonding energy of direct-bonded silicon wafers. Journal of Applied Physics 20039468006806
27. F Niklaus, G Stemme, J. Q Lu, R. J Gutmann, Adhesive wafer bonding. Journal of Applied Physics 2006
28. J Dlutowski, C. J Biver, W Wang, S Knighton, J Bumgarner, L Langebrake, W Moreno, A. M Cardenas-valencia, The development of BCB-sealed galvanic cells. Case study: aluminum-platinum cells activated with sodium hypochlorite electrolyte solution. Journal of Micromechanics and Microengineering 20071717371745
29. Y Okayama, K Nakahara, A Xavier, T Ninomiyia, Y Matsumoto, A Hotta, M Omiya, N Miki, Characterization of a bonding-in-liquid technique for liquid encapsulation into MEMS devices. Journal of Micromechanics and Microengineering 2010
30. Y Zhang, M Ishida, Y Kazoe, Y Sato, N Miki, Water vapour permeability control of PDMS by the dispersion of collagen poweder. TEEE:

Transactions on Electrical and Electronic Engineering 200943442449

31. S Sawano, K Naka, A Werber, H Zappe, S Konishi, Sealing method of PDMS as elastic material for MEMS. Proceedings of 21st IEEE International Conference on Micro Electro Mechanical Systems, MEMS 20081317January 2008Tuscon, USA.
32. M Antelius, A. C Fischer, F Niklaus, G Stemme, N Roxhed, Hermetic integration of liquids using high-speed stud bump bonding for cavity sealing at the wafer level. Journal of Micromechanics and Microengineering 2012
33. E. F Borra, O Seddiki, R Angel, D Eisenstein, P Hickson, K. R Seddon, S. P Worden, Deposition of metal films on an ionic liquid as a basis for a lunar telescope. Nature 20074477147979981
34. N Binh-khiem, K Matsumoto, I Shimoyama, Polymer thin film deposited on liquid for varifocal encapsulated liquid lenses. Applied Physics Letters 2008
35. N Binh-khiem, K Matsumoto, I Shimoyama, Tensile film stress of parylene deposited on liquid. Langmuir 201026241877118775
36. S Takamatsu, H Takano, N Binh-khiem, T Takahata, E Iwase, K Matsumoto, I Shimoayma, Liquid-phase packaging of a glucose oxidase solution with parylene direct encapsulation and an ultraviolet curing adhesive cover for glucose sensors. Sensors 201010658885898
37. Y Suzuki, Y. C Tai, Micromachines high-aspect-ratio paryelene spring and its application to low-frequency accelerometers. Journal of Microelectromechanical Systems 200615513641370
38. S Takeuchi, D Ziegler, Y Yoshida, K Mabuchi, T Suzuki, A parylene flexible neural probe integrated with micro fluidic channels. Lab on a Chip 20055519523
39. H Sasaki, H Onoe, H Osaki, R Kawano, S Takeuchi, Parylene-coating in PDMS microfluidic channels prevents the absorption of fluorescent dyes. Sensors and Actuators B: Chemical 20101501478482
40. J Watanabe, H Ishikawa, X Arouette, Y Matsumoto, N Miki, Demonstration of vibrational Braille code display using large displacement micro-electro-mechanical system actuators. Japanese Journal of Applied Physics 2012FL11.
41. O Regan, B Gratzel, M. A low cost, high-efficieny solar cell based on dye-sensitizd colloidal TiO2 films. Nature 1991
42. T Shigeoka, T Muro, T Ninomiya, N Miki, Wearable pupil position detection system utilizing dye-sensitized photovoltaic devise. Sensors and Actuators A: Physical 2008

43. A Oikawa, N Miki, MEMS-based eyeglass type wearable line-of-sight detection system. Proceedings of 2011 IEEE International Conference on Robotics and Automation, ICRA 2011913May 2011Shanghai, China.

Chapter 7

DESIGN AND FIELD TESTS OF A DIGITAL CONTROL SYSTEM TO DAMPING ELECTROMECHANICAL OSCILLATIONS BETWEEN LARGE DIESEL GENERATORS

Fabrício Gonzalez Nogueira[1], Anderson Roberto Barbosa de Moraes[1], Maria da Conceição Pereira Fonseca[1], Walter Barra Junior[1], Carlos Tavares da Costa Junior[1], José Augusto Lima Barreiros[1], José Adolfo da Silva Sena[2], Benedito das Graças Duarte Rodrigues[2] and Pedro Wenilton Barbosa Duarte[2]

[1]Federal University of Pará, Technology Institute, Faculty of Electrical Engineering, Belém, Pará, Brazil

[2]Northern Brazilian Electricity Generation and Transmission Company (ELETRONORTE- Eletrobrás), Brazil

INTRODUCTION

Electromechanical oscillations are natural phenomena in power systems having two or more synchronous generating units operating interconnected. These oscillations are undesirable because they can severely limit the power transfer between interconnected generating areas, due to reduced stability margins, as well as may decrease lifetime expectancy of system machines. If these electromechanical oscillations are not satisfactorily damped, they may, under some operating conditions, even increase (in amplitude), causing shutdown (tripping) of one or more interconnected generating units. As the oscillations are related to the physical nature of the electrical power system component's interactions, they cannot be avoided. However, by using efficient automatic control techniques, the electromechanical oscillations can be sufficiently attenuated in order assure a safe system operation, for all allowed operating conditions [1, 2].

Among the devices utilized to deal with electromechanical oscillations, the most common are the power systems stabilizers (PSS). As can be seen in Figure 1, dashed box (a), PSS devices usually actuates through the automatic voltage regulator (AVR) in order to increase the damping of poorly damped oscillations modes. The improvement of the damping is obtained through a torque component proportional to the machine speed deviation [1, 2].

An alternative technique, which has been investigated by several authors [3-5], is the application of a damping controller through the speed governor system (see dashed box (b), in Figure 1). This is the approach followed by this work. It is important to remark that this technique is recommended only for generators systems having fast response actuators, such as diesel engines. The installation of a damping controller via the speed regulation system may be advantageous because theoretical studies show that there is a weak coupling between this control loop and excitation system controllers in other machines of the interconnected power system [3, 4].

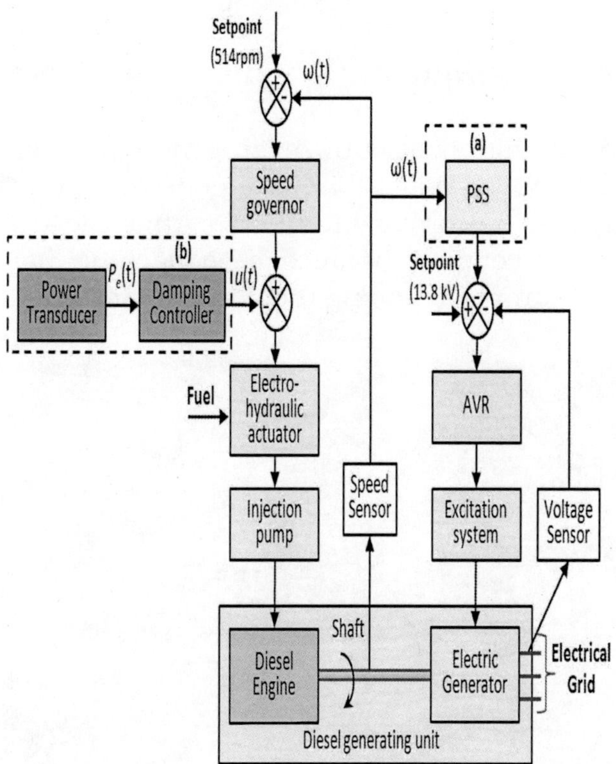

FIGURE 1. Comparison between a conventional PSS (a) and a damping controller actuating through the speed governor system (b).

A complete real world example of a control development for safe operation of a power plant has been presented in this chapter. The chapter details the design, implementation and field tests of a digital damping controller applied to a diesel generating unit, at Santana thermoelectric power plant (located in north of Brazil), addressing the system identification and the implementation of the damping controller using digital control techniques. The experimental tests results are presented and discussed.

SYSTEM DESCRIPTION

Diesel Generating Units At Santana Power Plant

Santana Power Plant has a set of seven generating units, including four identical 18-MVA Wärtsillä diesel generating units (Figure 2). Due to economic operation constraints, it is advisable to preferentially operating the diesel engines because these machines have the lowest specific fuel consumption among the power plant generating units.

FIGURE 2. Diesel generating units, at Santana Thermoelectric Power Plant.

The diesel generating units are driven by 18V46 Wärtsillä four-stroke diesel engines, which have 18 cylinders in V-form (V-angle of 45°), as shown in Figure 3. The engine presents a turbocharged and intercooled design, along with direct injection subsystem.

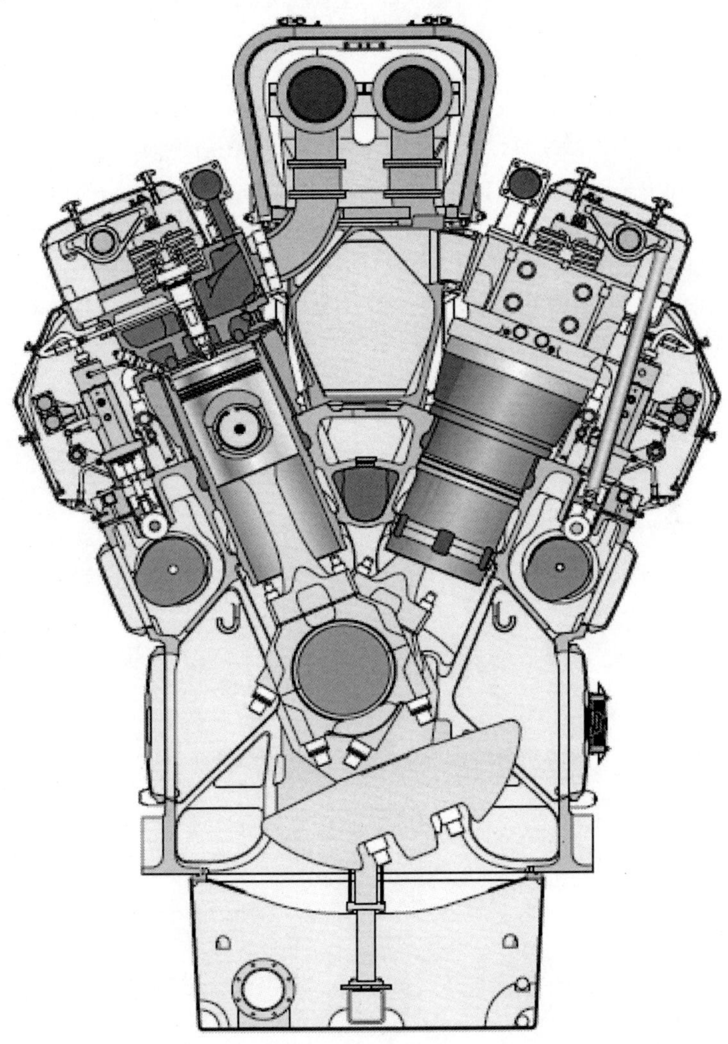

FIGURE 3. Cross section of the V-engine [6].

A simplified block diagram of the diesel generating units is shown in Figure 4. The turbocharging is performed by turbocompressors, which are driven by the exhaust gases from the combustion chamber. The turbine drives the compressor subsystem which in turn draws air from the environment increasing the air pressure. The compressed air is then cooled and, subsequently feedback into the combustion chamber, as illustrated in Figure 4. In order to keep the rotor speed at nominal value, during the system operation, a speed governor controls the loading of the diesel engine by actuating on the electro-hydraulic actuator position (see Figure 4). Therefore, more or less fuel is injected in order to increase or decrease the mechanical power demanded by the electrical generator and its load.

The damping controller proposed in this chapter actuates in the output of the speed governor, modulating the electro-hydraulic actuator position according to the observed electromechanical oscillation on the measured electric power. Table 1 shows the main technical characteristics of the 18V46 diesel engine:

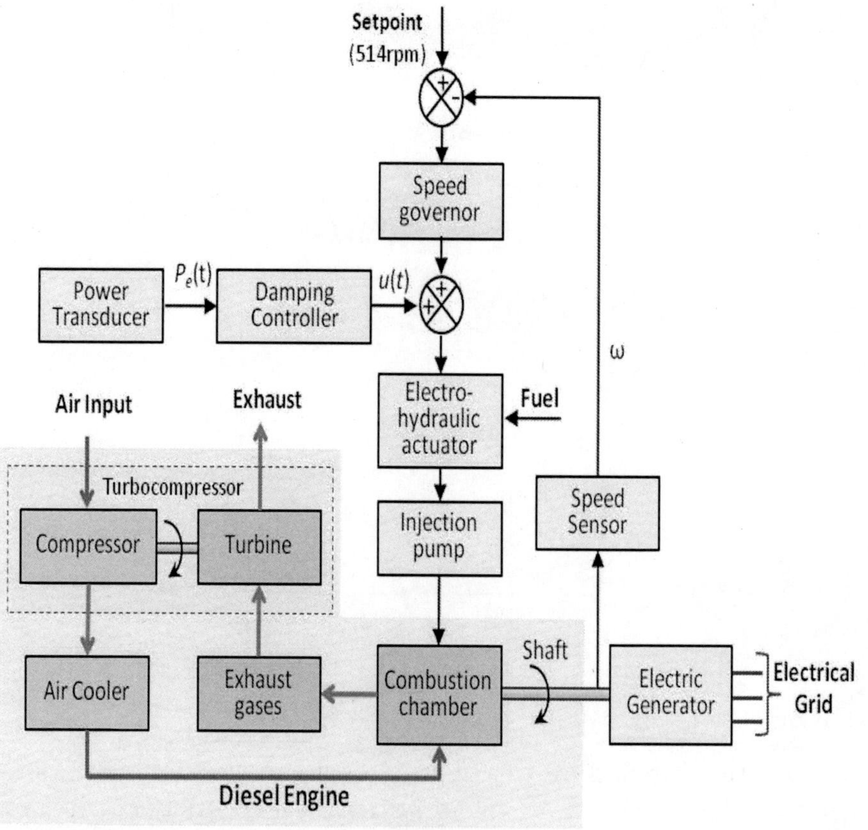

FIGURE 4. Simplified schematic of the diesel generating unit equipped with the damping controller.

TABLE 1. Main technical characteristics of the 18V46 diesel engine.

Main technical data of engine		
Parameter	**Unit**	**Value**
Engine output	kW	17,550
Generator output	kW	17,025
Specific consumption (Diesel oil)	Liter/kWh	0.22
Cylinder bore	mm	460
Piston stroke	mm	580
Cylinder output	kW/cyl.	975
Engine speed	rpm	514
Piston speed	m/s	9.9
Mean effective pressure	bar	23.6
Firing pressure	bar	180
Charge air pressure	bar	3.1
Weight (Engine + generator)	Ton	290

Depending on fuel availability, the 18V46 Wärtsillä diesel engines can be feed by one of the following fuel types: light fuel oil (light oil-diesel), heavy fuel oil, and natural gas. In Santana thermoelectric power plant, the light oil-diesel is the preferential choice. The light oil-diesel characteristics are presented in Table 2.

TABLE 2. Light oil-diesel characteristics.

Light oil-diesel characteristics		
Parameter	**Unit**	**Value**

Viscosity, max	cSt a 50°C	11
Density, max	g/ml	0.92
Sulphur, max	% mass	2
Vanadium, max	mg/kg	100
Ash, max	% mass	0.05
Water, max	% vol	0.3
Pour point	°C	6

The fuel injection system is composed of injection pumps, high pressure pipes, and injection valves. The nozzle is located at the top center of the cylinder head. The pressurized fuel in the low-pressure oil line (8.0 to 11.0 bar) has its flow controlled by the electro-hydraulic actuator, which is driven by the output signal of the speed governor (Figure 5). The fuel injected into the high-pressure combustion chamber (180 bar) is atomized and the combustion occurs after the compression cycle.

FIGURE 5. Electro-hydraulic actuator subsystem.

LOW-DAMPED ELECTROMECHANICAL OSCILLATIONS

Based on a priori information, provided by utility technical reports [7], a dominant 2.5 Hz intra-plant electromechanical oscillation mode was identified having a much reduced damping, which was observed from the measured machine terminal electrical power signal (Figure 6), when was applied a step in the electro-hydraulic actuator of the generating unit 2. As can be seen, diesel generating units 2 and 3 (G2 and G3) oscillated on phase opposition, which is interpreted as an indicative of intra-plant oscillation mode [2]. These tests were performed in order to investigate the reasons for the frequent machine shutdown (trip) due to actuating of torsional mode protection. Based on this a priori information, the 2.5 Hz intra-plant oscillation mode was then chosen as the target oscillation mode for which the digital controller presented in this chapter was designed.

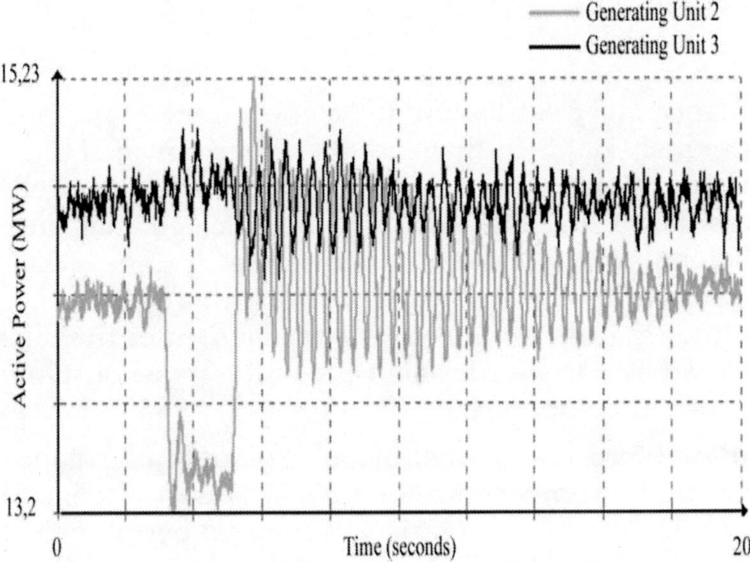

FIGURE 6. G2 and G3 response to a step variation applied to the G2 electro-hydraulic actuator [7, 8].

DIGITAL DAMPING CONTROLLER

In order to perform the field tests, the damping controller was implemented on an embedded system, composed by the main blocks: input conditioning system, digital controller, actuation system, local HMI, and communication interface (Figure 7).

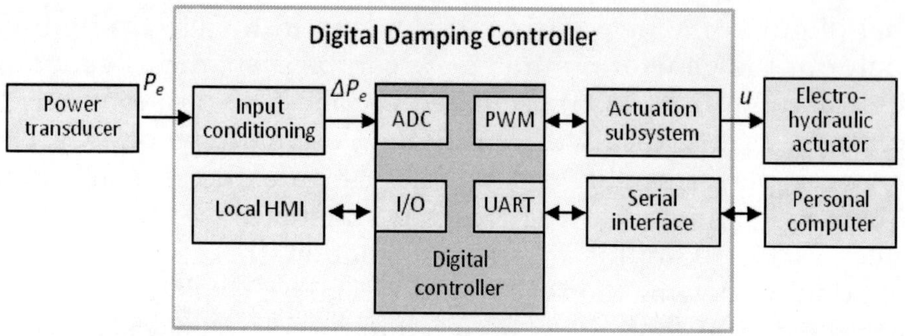

FIGURE 7. Damping controller blocks.

In order to obtain a signal having enough information about the dominant power oscillation, it was necessary to pre-processing the measurement data. To that end, the active power signal (P_e) has been chosen as the feedback signal to the damping controller. This choice was based on field tests, which indicated that the signal/noise ratio can be improved. Therefore, the P_e signal was applied in the conditioning module, where it was filtered by a first-order low-pass filter in order to attenuate noise and, after that, the signal was processed by a first-order high-pass filter (washout) in order to eliminate the DC component.

The digital processing module of the damping controller is based on a digital signal controller (DSC). This device incorporates features of microcontrollers (variety of internal peripherals) and the DSP's (specific support for digital signal processing). The analog signal ΔP_e is converted to digital through an internal 12-bits analog-to-digital converter (ADC). The firmware of the proposed damping controller control law was embedded in the digital signal controller and its control law was programmed by using C language.

The output of the digital controller is generated through an internal pulse width modulation (PWM) module of the DSC. The action of the damping controller in the electro-hydraulic actuator of the diesel engine is achieved by means of an actuator subsystem, which modulates the output current of the speed regulator. This electronic sub-system is implemented as an array of power transistors, which composes a current mirror scheme, which is commanded by the PWM signal.

The communication between the embedded system and other devices is performed through a RS-232 serial interface. Through this communication link, the embedded system is able to transmit the collected data to a personal computer (PC), to perform analyzing and control. The developed embedded system was designed to operate in four different operational modes, namely: (i) "step response", (ii) "identification", (iii) "control", and (iv) "configuration mode". When in the first mode, the equipment applies a step variation to the electro-hydraulic actuator, collect the plant response and send the collected data to an auxiliary microcomputer PC, for a more complex data analysis. When operating in identification mode, the damping controller may generate a pseudo-random binary sequence (PRBS) and uses this signal test to modulate the electro-hydraulic actuator position of the diesel engine. Using a non-recursive least mean squares algorithm, pairs of input and output data are used to obtain plant estimated parametric models, which are used on the controller design. When in "control mode", the damping controller acquires the signal of active power deviation and processes the damping control law, generating a control signal that is applied in the actuator of the fuel valve of the diesel motor, in order to damp the electromechanical oscillations. Finally, in "configuration mode", the user can set up configuration parameters of the controller through a keyboard and a menu on a LCD display (local HMI).

IDENTIFICATION OF LINEAR MODELS

Model Structure

In order to design the digital damping controller, it is necessary to estimate a mathematical model that represents the system dynamics at a specific operation condition. This modeling step can be performed using identification techniques, establishing a system dynamical model from measured input and output data, which are respectively, the current of command of the fuel valve and the signal proportional to the active power of the generator. The identified model captures the relevant information about the plant dynamic, for feedback control objectives.

A dynamic system model can be represented in several different ways, particularly in time and frequency domains. When the objective is to obtain a dynamic linear model around an operation point using sampled data, a discrete linear parametric model can be used such as an autoregressive with exogenous inputs (ARX) model. The ARX model can be represented in the discrete time domain by [9,10]:

$$A(q-1)y(k)=B(q-1)u(k)+v(k)$$

Or by the transfer function:

$$Gp(q-1)=B(q-1)/A(q-1)$$

where $q-1$ is the discrete-time delay operator, $y(k)$ and $u(k)$ are the sequences of input and output data, respectively, and $v(k)$ is assumed to be a white noise, a signal having a flat power spectral density, which is a very useful property for analysis and system identification [9, 10]. $B(q-1)$ and $A(q-1)$ are discrete-time polynomials in the form:

$$B(q-1)=b1q-1+b2q-2+\ldots+bnbq-nb$$

$$A(q-1)=1+a1q-1+a2q-2+\ldots+anaq-na$$

Coefficients $b1,\ldots,bnb$ and $a1,\ldots,ana$ are the model parameters, and n_b and n_a are integer numbers used to define the model order.

ESTIMATION ALGORITHM

The input-output (I/O) representation (1) can be put into a linear regression form as follows. Let us define a $n=na+nb$ vector which contains all the coefficients to be identified:

$$\theta T=[a1\ a2\ \ldots\ ana\ \ b1\ b2\ \ldots\ bnb]$$

We also define the extended regressor, which will be made up of past I/O data:

$\phi T(k)=[-y(k-1)\ \ -y(k-2)\ ...-y(k-na)\quad u(k-1)\ \ u(k-2)\ ...\ u(k-nb)\ ]$

The model output $y\hat{}(k)$ can be calculated through the product $\theta\hat{}T(k)$ $\varepsilon(k)$. Therefore, the estimating error $\varepsilon(k)$ is the difference between the measured system output $y(k)$ and the estimated output $y\hat{}(k)$, i.e., $\varepsilon(k)=y(k)-\theta\hat{}T(k)\phi(k)$.

The actualization of the model parameters $\theta\hat{}$ can be performed through an iterative algorithm, such as the least mean square method (LMS):

$$\varepsilon(k)=y(k)-\theta\hat{}T(k)\phi(k)$$

$$\theta\hat{}(k+1)=\theta\hat{}(k)+\gamma\varepsilon(k)\phi(k)$$

where the parameter γ is the step size.

PERSISTENCY OF EXCITATION AND DATA ACQUISITION

Before data acquisition, a previous system study is necessary to find the best way to obtain dynamic information of the plant. Analyzing system modes frequencies, an appropriate excitation test signal can be designed, which will excite the plant in a range of desired frequencies. This operation results in a better capture of dynamic information of the system, thus improving the dynamical model estimation.

The system to be identified has its dominant modes in a characteristic range of frequencies. Therefore, to excite the plant in an appropriate manner, the input signal must be designed to have an approximate uniform power spectrum in the dominant range. An exciting signal satisfying this property is the pseudo-random binary sequence (PRBS), which is a binary signal that can be generated by using digital techniques [10]. This can be done by using a feedback shift register having a length of N cells and a sample generation interval T_b. From knowledge of the minimum and maximum values f_{min} and f_{max} for the desired range of frequency excitation, the values of N and T_b can be calculated by using equations (9) and (10) [9, 10, 11]:

$$fmax=0{,}44Tb$$

$$fmin=1(2N-1)Tb$$

DESIGN OF THE CONTROLLER TO DAMP ELECTRO-MECHANICAL OSCILLATIONS

Pole Shifting Technique

The damping controller goal is to increase the dominant mode damping, without changing significantly the natural frequency (ω_n) of this mode. In order to perform this task, the pole shifting technique was utilized to obtain a controller able to provide a stable closed loop system and performance characteristics as specified in accordance with the designer requirements. This method is a particular case of the general pole placement method.

In this technique, the open-loop dominant poles must be radially shifted to a new position toward the origin of the unitary circle in the z-plane. The amount of the radial displacement is specified by a contraction factor α, according to the desired degree of damping for the closed loop system [12, 13]. Therefore, the designer first specifies a desired value, ξd, for the damping of the electromechanical mode and then calculates the value of the shifting factor α, by using:

$$a=e-(\xi d-\xi)\omega n Ts$$

where $0 \leq \alpha \leq 1$, Ts is the controller sampling period, ξ is the natural damping (system without damping controller).

The pole shifting method is based on the search of polynomials $R(q^{-1})$ and $S(q^{-1})$, which satisfy the polynomial equation (12), known as Diophantine equation:

$$A(q-1)S(q-1)+B(q-1)R(q-1)=Acl(q-1)$$

Where the polynomials A(q−1) and B(q−1) are both known for the designer (model estimated of the plant), Acl(q−1) is a polynomial with the desired closed-loop poles, and R(q−1) and S(q−1)are the polynomials with the parameters of the controller to be calculated:

$$R(q-1)=r0+r1q-1+r2q-2+\ldots+rnrq-nr$$

$$S(q-1)=1+s1q-1+s2q-2+\ldots+snsq-ns$$

Assuming that na = nb = n and nr = ns = n – 1 (minimum order controller), and matching the coefficients of the same power in q−1, the coefficients of the control law can be obtained directly by the solution of the following linear equation system:

[10.0b10.0a11.0b2b1.0.a1...b2..ana..1bnb..b10ana.a10bnb.b2.0.0.0..........00.

ana00.bnb][s1..snsr0..rnr]=[(α−1)a1(α2−1)a2..(αna−1)ana0.0]

SMITH PREDICTOR

Field tests performed at the diesel power plant revealed that there is a considerable dead time. For the tests described in this work, dead time is the time delay taken for the electric power signal starts to react after the application of a variation on the electro-hydraulic actuator. Smith Predictor is an adequate method to design controllers taking into account the observed delay [13]. The resulting controller with Smith Predictor consists of the digital controller C(q−1)=R(q−1) /S(q−1) with two additional internal feedback loops (see Figure 8), where one is a linearized estimated model of the plant without considering the time delay G^p(q−1), while the other takes into account the model with the time delay q−d^G^p(q−1).

EXPERIMENTAL FIELD TESTS

The damping controller was successfully installed and validated by experimental field tests carried out at Santana Power Plant. The controller equipment has been installed in a cabinet of the control system of an 18 MVA diesel generating unit (Figure 9). Tests for model identification and estimation, with subsequent design and test of the proposed damping controller operating in closed-loop have been performed, with results showing an increase of the damping of the dominant electromechanical mode.

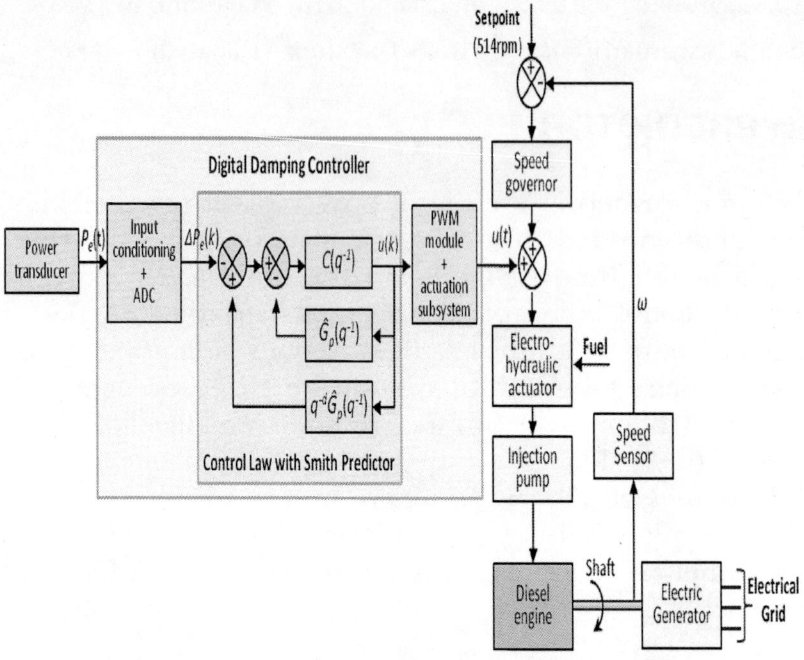

FIGURE 8. Digital controller with Smith Predictor.

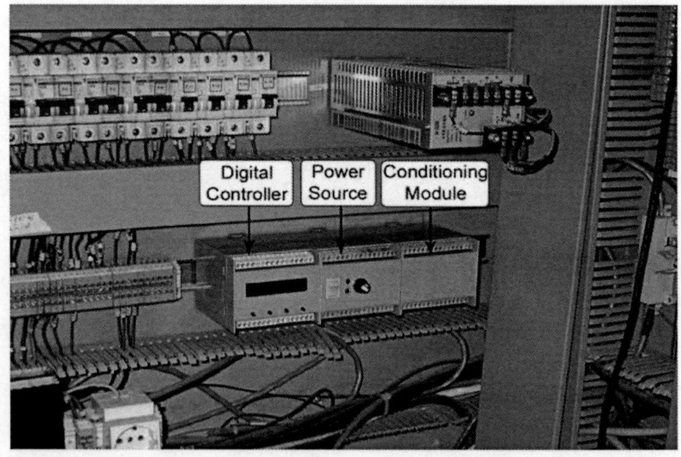

FIGURE 9. Field tests of the damping controller prototype, at Santana power plant [8].

ANALYSIS OF THE SYSTEM RESPONSE FOR THE APPLICATION OF A STEP VARIATION

Step response is a useful way for an initial understanding of the system dynamic behavior, revealing some important characteristics that can be used for system modeling. Thus, with the damping controller disabled and the developed equipment programmed to operate only as a step generator, step variations were applied in the command of the fuel valve of the diesel engine. In consequence, the dominant oscillation mode with a frequency around 2.5 Hz, was observed in the electrical power signal. The amplitude of the test signal was configured to 5 mA, which is equivalent to an increase of 5% in the steady state opening of the fuel valve at the operation point considered, while the steady-state valve opening is set at 50 % (9 MW of active power).

These initial tests were also useful to identify possible nonlinearities of the system, such as deadband and time delay. These phenomena are usually found in hydraulic and thermal systems, such as the electro-hydraulic actuator and the diesel engine, and if they are ignored, may be difficult to tune the control system. In order to assess the nonlinearities, a series of tests was performed, which showed that the electro-hydraulic actuator has an excellent sensitivity and a dead zone that does not compromise the system control. So it was not necessary to implement any deadband compensation strategy in the controller. The field tests have shown that the diesel engine actuation system presents a dead time around 400 ms. In order to deal with this observed dead time, a Smith Predictor strategy was applied, as described in Section 5.2 of this chapter.

SYSTEM IDENTIFICATION TESTS

With the goal of obtaining a parametric model for the damping controller design, identification tests were performed in the 18-MVA diesel generating unit without using a damping controller. In order to excite the system electromechanical modes of interest, a PRBS sequence was used, with the parameter *Tb* equal to 80 ms and *N* equal to 9, resulting in a minimum frequency of 0.02 Hz and maximum frequency of 5.5 Hz, which excite uniformly the range of possible frequencies of the electromechanical oscillations modes (between 0.2 to 3 Hz) [1, 2]. The point of application of the exciting PRBS signal is the same point used for the application of the step response test, as already described in Section 6.1.

The input and output data sets were collected with a 40 ms sampling period and automatically transmitted to a PC, in which data sets were processed for purposes of identifying models representing the system dynamics in the current operating point. Figures 10a and 10b illustrate, respectively, the data obtained from the plant input variable (current variation in the diesel engine valve admission control) as well as from the plant output variable (power generator active power deviation).

From the acquired data, it is possible to make an estimate of the system response frequency spectrum on the operating point considered. The PRBS spectrum is characterized by being approximately uniform in the range specified by the project (Figure 10c), meaning that all modes between 0.1 and 5 Hz (approximately) were equally excited by the designed test signal. As can be seen in Figure 10d, a 2.5 Hz dominant intra-plant mode, having a small damping, can be observed from the electrical power deviation signal. Therefore, it is advisable to design a damping controller in order to improve the damping of this dominant oscillation mode.

The acquired data was divided into two data sets. The first one was used for the parametric model identification process, while the other set was used for model validating purposes. The identification process was carried out using a non-recursive least squares algorithm [10]. A fourth order ARX model structure was chosen, having 4 parameters in the numerator (B), 4 parameters in the denominator (A) and a discrete-time delay of 10 sampling intervals (400 ms).

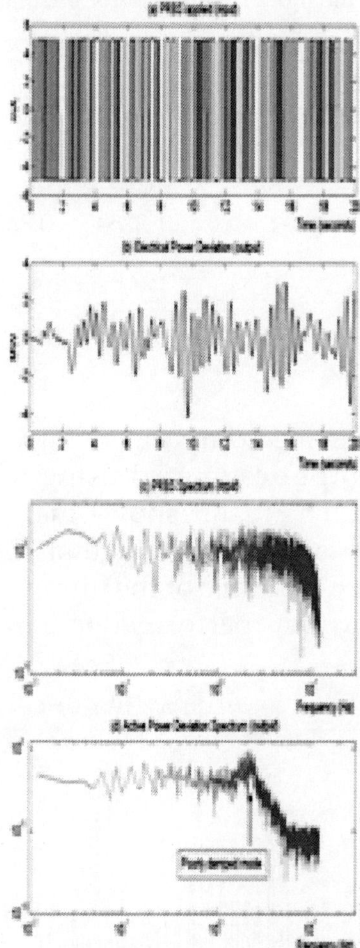

FIGURE 10. a) PRBS applied in generator (b) Active power deviation of the generator (c) PRBS spectrum (d) Active power deviation spectrum [8].

In this way, the resulting identified plant input-output ARX model, which was used for control design, has the following structure:

$$q^{-d}\hat{G}_p(q^{-1}) = q^{-9}\frac{B(q^{-1})}{A(q^{-1})} = q^{-9}\frac{b_1q^{-1}+b_2q^{-2}+b_3q^{-3}+b_4q^{-4}}{1+a_1q^{-1}+a_2q^{-2}+a_3q^{-3}+a_4q^{-4}}$$

where the parameters values are presented in the Table 3.

TABLE 3. Parameters of the model estimated for the plant ($Ts = 0.04$ s, $d = 10$ sampling intervals).

Parameter	**a1**	**a2**	**a 3**	**a 4**
Value	-1.9980	1.8254	-0.8676	0.2626
Parameter	**b1**	**b2**	**b3**	**b4**
Value	0.0033	0.0034	0.0030	0.0028

Figure 11 illustrates the comparison between the real output of the system and the output estimated using the fourth order model identified in the tests. The result shows the good estimation of the model parameters. It is also verified that the model successfully captured the oscillatory dynamic of the intra-plant electromechanical mode, observed in field tests performed in the diesel generating unit.

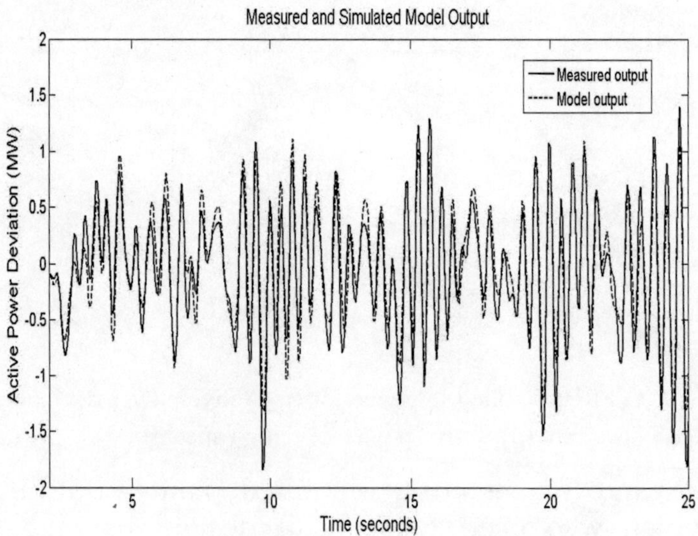

FIGURE 11. Comparison between the measured output (solid black) and the output of the model (dashed blue), for the diesel generating unit of 18 MW [8].

DESIGN OF THE DAMPING CONTROLLER CONTROL LAW

Using the identified model, the damping part of the control law was obtained using the pole shifting method and solving the matrix equation system (15). The controller design main goal is to provide an acceptable damping (in this case, $\zeta_d = 0.2$), without affecting substantially the electromechanical dominant mode natural frequency and not exciting too much any unmodelled dynamics. In the plant fourth order ARX model (B / A), it is considered only one delay of sampling interval. The additional delay of 9 sampling intervals (d=1+9) is accommodated through the control law scheme implemented using the Smith Predictor, as explained in Section 5. Table 4 shows the parameters of the damping controller used in field tests.

Parameter	r0	r1	r 2	r3
Value	-1.6484	0.5728	-3.7522	-0.3732
Parameter	s1	s2	s3	-
Value	0.1018	0.0287	0.0020	-

TABLE 4. Parameters of the designed digital damping controller (Ts = 0.04 s).

CLOSED-LOOP CONTROL TESTS

After the controller design, its parameters were inserted into the embedded controller. Then the device was programmed to act in the closed loop control mode (damping controller mode), to provide an additional damping signal to the system, through the speed control loop of the diesel generating unit. To evaluate the performance of the damping controller, step type disturbances were applied, as can be observed during the tests illustrated in Figure 12.

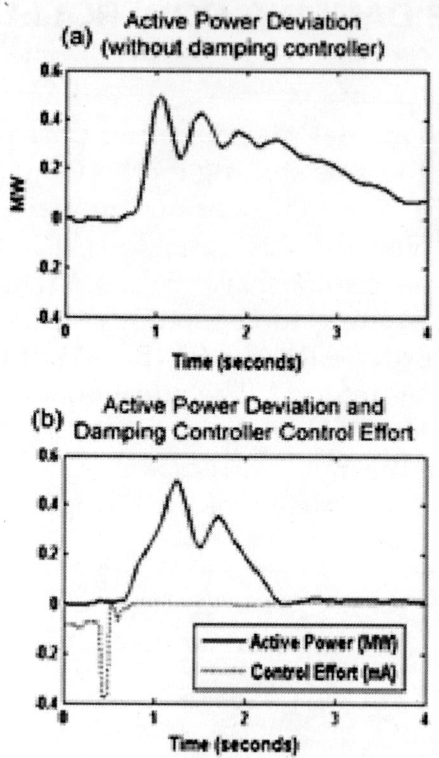

FIGURE 12. System step responses without damping controller (a) and with damping controller (b) [8].

Analyzing the graphics on Figure 13, it can be observed that the inclusion of the damping controller in the speed loop of Wärtsilä considerably improves the stability of the system, making it less oscillatory without affecting the speed governor performance.

In Figure 13 a comparison between the power spectral density of the active power deviation signal collected with the system with and without damping controller is shown. When the data were collected, the generator was excited by a PRBS signal. It is clear from this measurement that the observed electromechanical mode, around 2.5 Hz, is much more pronounced for the case in which the system operates without damping controller, than when the damping controller is acting on the system.

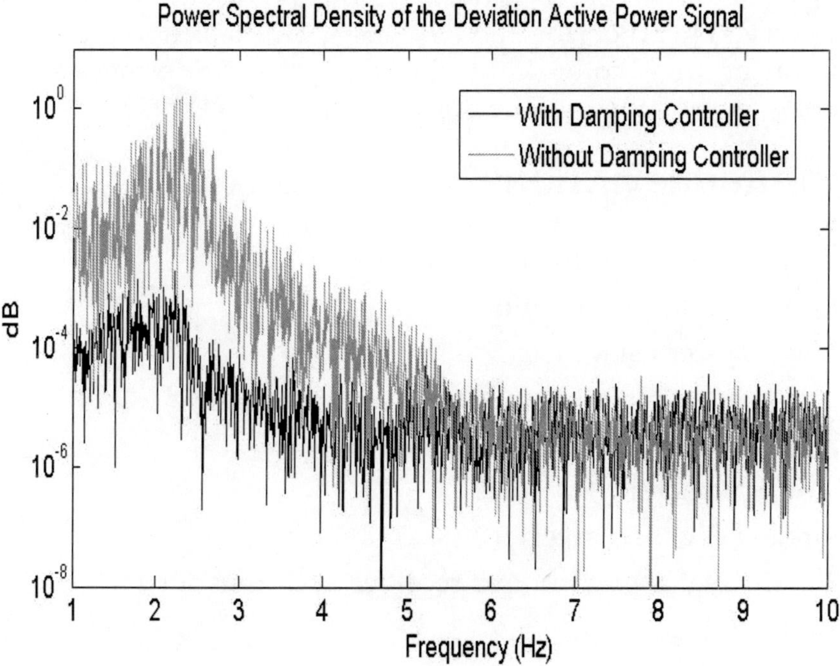

FIGURE 13. Comparison between the power spectral density of the deviation active power signals collected with damping controller and without damping controller [8].

CONCLUSION

This chapter presented the design and experimental tests of a digital damping controller which actuates through the speed governor system in order to damp an intra-plant electromechanical oscillation mode. The controller field tests were performed in a real generating unit of 18 MW in the Santana Thermoelectric Power Station.

The actuation of the damping controller through the speed governor was only possible due to the fast response of the diesel engine. Thus, this technology is not recommended for use in systems with machines that have low speed of actuation, as in generating units driven by hydraulic turbines.

Results demonstrated that the system performance was improved after the inclusion of the damping controller, thus increasing the system dynamic stability margins. No adverse interactions have been observed in all performed field tests. This task and others, such as, studies of an adaptive controller design, considering a variable time delay in the loop, will be object of future investigations.

LIST OF ABBREVIATIONS

ARX autoregressive with exogenous input

PWM pulse width modulation

PSS power system stabilizer

AVR automatic voltage regulator

G_I i-th generating unit

HMI human machine interface

ADC analog digital converter

UART universal asynchronous receiver and transmitter

I/O input/output

DC direct current

DSC digital signal controller

DSP digital signal processor

PC personal computer

PRBS pseudo random binary sequence

LMS least mean squares

LIST OF SYMBOLS

Ts data acquisition and control sample time

$q-1$ discrete-time delay operator ($q-1y(t)=y(t-1)$)

k normalized discrete time

$B(q-1)$, $A(q-1)$numerator and denominator polynomials of the plant ARX model θvector of parameters□data regressor vectorεestimating errorγstep size of LMS algorithm$R(q-1)$, $S(q-1)$numerator and denominator polynomials of the digital controller

N, Tb number of cells and sample generation interval for the shift register

f_{max}, f_{min} maximum and minimal values for the desired excited frequency range

ωn dominant oscillation mode natural frequency

$Acl(q-1)$ specified closed-loop polynomialξd specified value of the damping for the dominant electromechanical modeξestimated value of the damping for the dominant electromechanical mode

α pole-shifting factor

$G\hat{}p(q-1)$estimated model of the plant without considering the time delay$d\hat{}$estimated process discrete-time delay$q-d\hat{}G\hat{}p(q-1)$estimated model of the plant considering the time delay

P_e active power signal

u control signal

ω rotor speed

ACKNOWLEDGEMENT

The authors acknowledge the support from ELETRONORTE (Northern Brazilian Electricity Generation and Transmission Company), through the R&D project number 4500049067 (2005), and from CNPq (The Brazilian National Council of Research).

REFERENCES

1. Prabha S Kundur (1994Power System Stability and ControlMcGraw-Hill.
2. Graham Rogers (2000Power System OscillationsKluwer Academic Publishers Group.
3. H F Wang, F J Swift, Y S Hao, B W Hogg, 1993Stabilization of Power Systems by Governor-Turbine ControlElectrical Power & Energy Systems, 156
4. H F Wang, F J Swift, Y S Hao, and B W Hogg, 1996Adaptive Stabilization of Power Systems by Governor-Turbine ControlElectrical Power & Energy Systems, 182
5. S K Yee, J V Milanović, F M Hughes, 2010Damping of system oscillatory modes by a phase compensated gas turbine governorElectric Power Systems Research80667 EOF674 EOF

6. Wärtsilä (2007Project guide Wärtsilä 46Finland.
7. Eletronorte2000Field Tests to Perform Adjusts in the Parameters of the Speed Governor Controllers of the Santana Power Plant- Wärtsilä Generating Units. Technical Report (in Portuguese).
8. F G Nogueira, J Barreiros, A L Barra, Jr. W, Costa Jr. C T, Ferreira A M D. (2011Development and Field Tests of a Damping Controller to Mitigate Electromechanical Oscillations on Large Diesel Generating UnitsElectric Power Systems Research812725732
9. D Ioan, Landau, Gianluca Zito (2006Digital Control Systems: Design, Identification and Implementation.ed. Springer.
10. Lennart Ljung1987System identification: Theory for the user" University of Linköping Sweden, Prentice Hall, Englewood Cliffs, New Jersey, 1987.
11. Paul HorowitzWinfield Hill (1989The Art of ElectronicsCambridge University Press, New York, USA, 2nd Edition.
12. A José, L Barreiros (1989A Pole-Shifting Self Tuning Power System Stabilizer. MSc Thesis, UMIST, Manchester- UK.
13. J Karl, ström, Bjorn Wittenmark (1997Computer Controlled Systems-Theory and Design, 3rd Edition, Prentice-Hall.

Chapter 8

THE GEOMECHANICAL INTERACTION OF MULTIPLE HYDRAULIC FRACTURES IN HORIZONTAL WELLS

Sau-Wai Wong[1], Mikhail Geilikman[1] and Guanshui Xu[2]

[1] Shell Exploration and Production Inc., Houston, Texas, USA

[2] FrackOptima Inc., and The University of California Riverside, California, USA

The technology of multiple hydraulic fracture stimulation in horizontal wells has transformed the business of oil and gas exploitation from extremely tight, unconventional hydrocarbon bearing rock formations. The fracture stimulation process typically involves placing multiple fractures stage-by-stage along the horizontal well using diverse well completion technologies. The effective design of such massive fracture stimulation requires an understanding of how multiple hydraulic fractures would grow and interact with each other in heterogeneous formations. This is especially challenging as the interaction of these fractures are subject to the dynamic process of subsurface geomechanical stress changes induced by the fracture treatment itself. This paper consists of two parts. Firstly, an idealized analytical model is used to highlight some key features of multiple hydraulic fractures interaction, and to provide a quantification of 'stress shadow'. Secondly, a new non-planar three dimensional (3D) hydraulic fracturing numerical model is used to provide an insight into the growth of multiple fractures under the influence of subsurface geomechanical stress shadows. Attention is given to studying the height growth of multiple fractures.

INTRODUCTION

Hydraulic fracturing or fracture stimulation establishes conductive fractures hydraulically from a horizontal well in the tight formation/ reservoir. They provide large surface area contact with the formation and thus facilitate the production of oil and gas, as evident from the experience in North Americas [1]. Multiple fractures are now placed in sequential stages in horizontal wells. Typically, four or more fractures are pumped simultaneously into a single frac stage, and it is not uncommon to place 20 to 40 frac stages in a single horizontal well. Attempts have also been made to simultaneously fracture two adjacent horizontal wells to generate complex fracture networks thought to be beneficial for production [2, 3].

It is important to consider the changes to subsurface in-situ stresses induced by fracture stimulation. This fracturing induced formation stress perturbation is also called 'stress shadowing' and needs to be understood and quantified for optimization and to avoid problems such as fracture 'screen-out'. Added complexity is caused by the fact that the fractures are created in an extremely tight formation - where the fracture pumping operations are often conducted within a time span that is in the same order of magnitude needed for frac fluid to leak-off and the resultant stress perturbation to completely dissipate. Moreover, the stress perturbation caused by inclusion of millions of pounds of prop pant in the formation does not dissipate.

Observations from substantial amounts of field data show that generally 25% or more of the fractures placed are ineffective. In fact, recent fracture stimulation surveillance with advance technologies such as microseismic monitoring and fibre optics temperature and strain sensing [4-6] have shown that multiple fractures do not grow and develop in the same way. Some of the fractures are prematurely 'terminated' during the treatment. Nonetheless, to increase production, there is a tendency to use longer horizontal wells, place more fractures per stage, longer fractures and more stages of fractures. Consequently, more fractures with closer spacing and larger volumes of frac fluid are employed. This translates to higher pumping rates, higher treating pressure and surface horse power, and consumes more materials (frac fluid, chemicals and proppant). The optimal design of hydraulic fractures clearly necessitates an engineering optimization, considering production, treatment cost and the feasibility of placing these closely spaced conductive fractures in the subsurface. The latter requires knowledge of key parameters such as in-situ stress, rock stiffness and strength, frac fluid rheology and leak-off behaviour, and importantly, the impact of stress shadows on multiple fracture growth in heterogeneous formations [7-13]. Some experts have aimed to provide general guidelines for the design of optimal fracture placement based on simplified analytical models whilst others have developed numerical models of various degree of sophistication to account for multiple fracture interaction and design [14-18].

In-situ stress perturbation can potentially result in the growth of complex non-planar fractures. An analytical model is developed to highlight some of the salient features of multiple non-planar hydraulic fractures interaction. The benefit of using an analytical model is that it provides immediate insights into the controlling parameters on fractures interaction, and guides further numerical analysis for stimulation optimization. Here, we present only the impact of fracture spacing and in-situ stress difference on fracture growth pattern. Other parameters that affects fracture growth pattern are evident from the model, but they will not be described here.

Emboldened by the results of our analytical model, a new numerical model based on rigorous mechanistic formulation has been devised to allow for more realistic solution of field problems. This is a non-planar 3D numerical model, where the interaction of multiple non-planar fractures is captured meticulously by means of boundary integral formulation with dislocation segments solution techniques. In the model, fracture growth and fluid flow equations are solved in a coupled manner via a proprietary, robust and efficient algorithm where mass conservation is strictly observed. This new non-planar 3D model was first described in [19]. It substantially enhances the capability of a plane strain 2D model presented in [7]. The present paper complements the published results [7, 19] and extends to considering fluid viscosity and height growth of multiple fractures.

ANALYTICAL MODEL

The complete theory of fracture interaction for an arbitrary number of fractures is quite cumbersome. Therefore, a 'simplified' analytical model has been developed to describe a set of fractures growing in a viscous mass-transfer dominated regime. In this model, two types of fracture growth patterns are defined. First is the 'compact' fracture growth where interaction and interference between fractures are minimum.

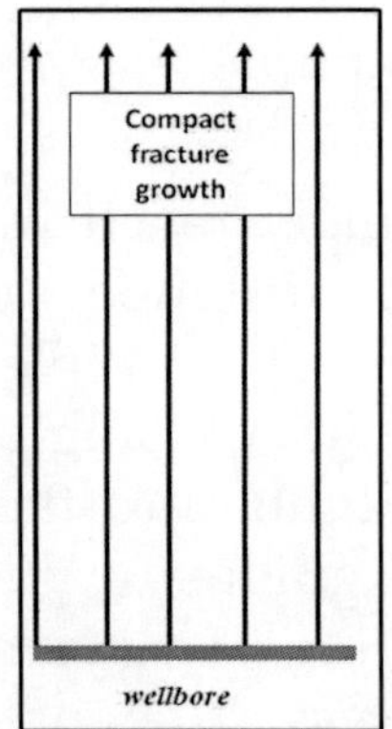

Weak fracture interference: all hydraulic fractures grow without interference

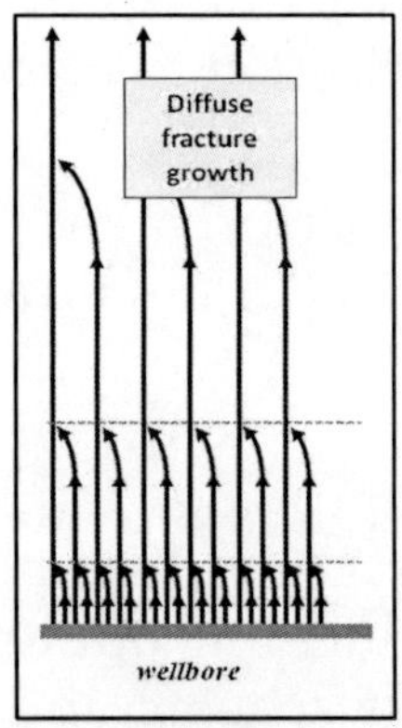

Strong interference: hydraulic fractures rotate and coalesce, or termination of some fractures

FIGURE 1. "Compact" and "Diffuse" fracture growth patterns

The second is 'diffuse' fracture growth where some fractures will either terminate or collapse into each other because of minimum in-situ stress rotation (Figure 1). In diffuse growth, we distinguish different zones. The first zone is the closest to the well. The next zone, farther from the well, contain less fractures because some of them have either terminated or merged into other fractures, as shown at the right-hand side of Figure 1.

The transition from one growth pattern to another is delineated by a 'critical' fracture spacing, influenced by the original principal stress difference in the un-fractured formation. In the case of diffuse growth, different zones are depicted by the dashed lines on the right-hand picture of Figure 1. The Average permeability of the fracture set in the zone "n" is given by (see e.g. [20]):

$$k_n = \frac{w_n^{\ 3}}{12 L_n} \tag{1}$$

where L_n is spacing between the fractures in the "n" zone, and average aperture of fractures in the zone "n" is:

$$w_n = a \frac{p_n}{E} R_n \tag{2}$$

Here, p_n is net pressure, E is Young's modulus, R_n is a radius of a penny-shape fracture, and a is a numerical coefficient (of approximately 1) and depends on Poisson's ratio. For simplicity, it is assumed that there is no leak-off of fluid into the extremely tight formation. Therefore, the average fluid density, m, in the fractures becomes:

$$m = \frac{w_n \rho}{L_n} \tag{3}$$

Where, ρ is fluid density. Applying the equation of fluid mass balance:

$$\frac{\partial m}{\partial t} + div\, \rho \vec{v} = 0 \tag{4}$$

and Darcy's law:

$$\vec{v}_n = -\frac{w_n^3}{12\mu L_n} \cdot grad(p_n) \tag{5}$$

We obtain an equation for fluid transport in the fractures:

$$\frac{\partial}{\partial t}(\frac{w_n}{L_n}) - \frac{E}{12\mu a} div\{\frac{w_n^3}{L_n} \cdot grad(\frac{w_n}{R_n})\} = 0 \tag{6}$$

The above Eq. 6 is highly nonlinear and cannot be solved analytically, instead, it is solved by considering the integral relationships and using a modification of the so-called method of successive change of steady states [21]. The essential idea is to solve the steady-state version of Eq. 6 for a given size of the fracture (i.e. without the first term in Eq. 6).

Subsequently, the time dependency of fracture radius growth is found from the integral form of mass balance equation, the differential form of which is given by Eqs. 4 and 6. The detail of this solution will be given in a future paper.

Interaction between fractures is governed by the induced stress due to the growing fractures. Solutions for various cases of interacting fractures have been published [22]. Since these solutions are cumbersome, several approximations have been proposed [23], which make the problem more tractable for our present purpose. Following these approximations we can obtain the stress state induced by two neighboring growing fractures. The induced stress depends on the fracture aperture (or equivalently, net pressure) and a dimensionless ratio of fracture length and the distance between fractures (i.e. the initial fracture spacing).

To appreciate the geomechanical interaction of fractures, see the left side of Figure 2, which depicts a top view of two simultaneously growing fractures. We choose to examine the incremental stress changes in the x and y directions at a mid-point in between the fractures, as shown in the figure. The induced stress is obviously not constant in the space between the fractures. But for illustration purpose, we will simply consider the stress evolution at this chosen point to be representative of the stress state of the large area between the fractures.

The analytical model on the right side of Figure 2 shows that σ_x increases with fracture length. Although σ_y also increases, it does so at a much slower pace than σ_x. We can now define a deviatoric stress, $\Delta\sigma = \sigma_x - \sigma_y$, which exhibits a maximum value. It can be immediately observed that if the maximum value of $\Delta\sigma$ is greater than the original in-situ stress difference ($\sigma_H - \sigma_h$), then rotation of minimum in-situ stress direction is feasible and the fracture(s) would grow in a non-planar fashion. For multiple fracture growth, this will result in the collapse or termination of some fractures.

In the same way, the height growth of multiple fractures may also see similar interaction if $\Delta\sigma$ is greater than the original in-situ stress difference between vertical stress and minimum horizontal stress ($\sigma_V - \sigma_h$). As fractures grow upward, assuming no stress barrier or contrast is encountered, both σ_V and σ_h would typically decrease in magnitude. Potentially, the stress shadowing effect is even more prominent for height growth of multiple fractures. It is interesting to note that field experience does indicate that it is challenging to grow in fracture height for multiple fracture stimulation. A plausible explanation is indeed the 'stress shadow' effect caused by geomechanical interaction of multiple fractures.

Although the analytical solution is an approximation, it captures the foremost qualitative features of fracture interaction. One of the first results shows how the in-situ stress contrast influences the domains of fracture growth.

Figure 3 shows the results based on the analytical model, where a distinct transition between the two regimes of fracture growth can be identified. Note that in practice, typical fracture spacing is in the order of 70 to 100ft (20-30m). Therefore, according to this simplistic, homogeneous model, today's multiple hydraulic fracture stimulation practices can potentially experience 'strong interference', depending on the stress contrast.

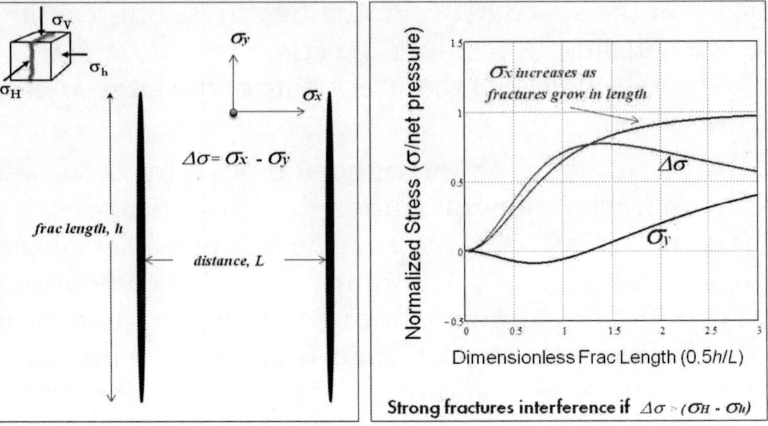

FIGURE 2. Interaction between two fractures

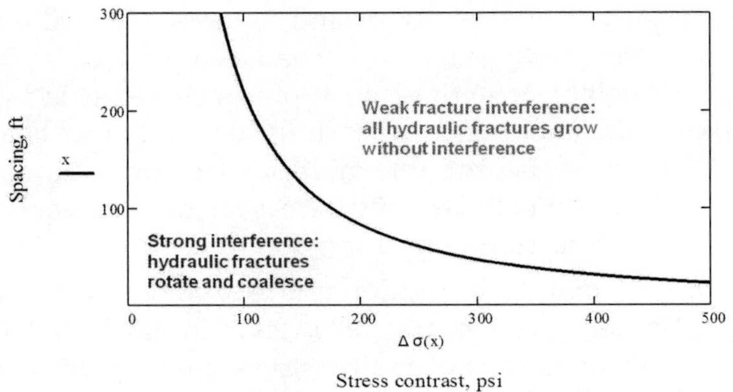

FIGURE 3. Fracture spacing versus stress contrast

NON-PLANAR 3D MODEL OF HYDRAULIC FRACTURES

The numerical simulator essentially follows a well-accepted numerical methodology [24, 25]. In this numerical framework, the rock mechanics of fracture growth takes place in a linear elastic medium and therefore can be modelled by boundary integral equations based on fundamental solutions of dislocation segments. The fracture growth is controlled by both fracture mechanics criteria and fluid volume balance at the fracture tip. The formulation adequately portrays both viscosity and toughness controlled fracture growth as well as the 'fluid lag' behavior [25]. The prop pant laden frac fluid is idealized as an incompressible power-law fluid commonly characterized by parameters K and n:

$$\tau = K\dot{\gamma}^{n} \tag{7}$$

Where τ is the fluid shear stress and $\gamma^{\cdot}$ the fluid shearing rate. K and n are termed the consistency index and power law index, respectively. They are naturally dependent on the prop pant concentration, and this effect of prop pant concentration on fluid rheology is accounted for in the simulator using Shah's model [27]. The prop pant transport and settlement are computed via the transport equation and empirical equations of particles settling between parallel plates [28]. The fluid loss to the formation is described by a Carter type leak-off model [25].

The ensuing flow of prop pant laden frac fluid within a fracture is captured according to the Reynolds lubrication theory, and the derived non-linear fracture growth and fluid flow equations are now solved in a coupled manner in which mass conservation (i.e., frac fluid and prop pant) are strictly enforced. The equations are discretized on the same structured grid, and they form a system of moving boundary, transient coupled equations of fracture width and fluid pressure. A new adaptive time integration algorithm has been developed to provide numerically stable solutions for a wide range of power-law fluid properties (especially when the viscosity is low).

Although the prop pant transport is calculated at each time step based on a volume conservation equation, this computation is decoupled from the process of solving the fluid flow equation for the sake of computational efficiency. Specifically, the fluid pressure and fracture width are first obtained by 'freezing' the prop pant concentration associated with each element. The contribution of prop pant transport to the volume concentration is then updated based on the calculated fluid velocity field. This proves to be effective as well as efficient since the employed time step is generally small.

NON-PLANAR FRACTURE PROPAGATION

In dealing with non-planar fracture growth, the heterogeneous formation is modelled as multiple isotropic parallel layers with heterogeneities characterized by variations in in-situ stress, elastic stiffness modulus, Poisson's ratio, fracture toughness, pore pressure, leak-off coefficients, and spurt-loss coefficients. All fractures are assumed to be vertical but they can turn in any horizontal direction. For each fracture, the growth direction is determined by the stress state of the leading element at the fracture tip. These assumptions exclude the application for certain geological formations. In particular, geological stress regimes as a result of reverse faulting may present difficulties in modelling as the hydraulic fracture would most likely grow in a horizontal plane. However, the assumptions made are sufficient and adequate for typical deep unconventional resource formations where the vertical stress is generally the maximum principal stress, or at least greater than the minimum horizontal principal stress – resulting in vertical fracture growth.

Fractures advance when the maximum tensile stress ahead of the crack tip exceeds the intrinsic tensile strength of the rock. In linear elastic fracture mechanics, the stress at the crack tip is singular, therefore the stress intensity factor, which correlates to the strength of such a singular stress field, is generally used to determine fracture tip propagation [29]. In practice, hydraulic fracturing processes take place under relatively large compressive in-situ earth stress that is normally several orders of magnitude higher than the rock strength. Therefore, the process is dominated by the balance of the compressive stress and fluid pressure at the fracture tip region. A group of 'virtual' elements is placed along the crack front and at each time step a check is made on the stress status of these elements. The virtual elements are allowed to be active as part of the new fracture surface if the potential fluid pressure can overcome the compressive stress plus the strength of the rock.

Importantly, the interaction of multiple fractures may result in significant shearing displacements along the fracture surface, causing the fracture growth to follow a curved path, and the computational technique leads these virtual elements to be oriented according to the stress field so that the local minimum principal stress is perpendicular to the new fracture surface as the fracture front advances.

An outline of the mathematical description of the fracture stiffness matrix, the displacement discontinuity method is given in [19], and will not be repeated here. The numerical approach is robust and efficient but less accurate than the variational boundary integral method which employs second order elements [30, 31]. For the current application, computational efficiency is deemed to outweigh the importance of a moderate improvement in accuracy. [19] also describes the coupling of derived non-linear fracture growth to flow equation and mass conservation.

WELLBORE HYDRAULIC MODEL

The fractures are assumed to be initiated at the specified locations of a cluster of perforations. Near wellbore formation stress concentration is not considered. The downhole pressure is defined as the fluid pressure just upstream of all injection points. It is obvious that the distribution of the corresponding injection at each fracture follows wellbore fluid mechanics, depending on the injection area connected with the fracture and the fluid pressure within the fracture. An approximation is made by using Bernoulli equation to capture the fluid distribution. This wellbore hydraulic model is able to account directly for the empirically derived or calibrated frictional pressure drop at the perforation and along the wellbore. Therefore, it is possible to account for limited entry perforation designs commonly employed in multiple hydraulic fracture design.

COMPARING SIMULATION RESULTS WITH ANALYTICAL SOLUTION AND EXPERIMENT

The complete validation of a hydraulic fracturing simulator is enormously challenging, because of the difficulty in obtaining the general analytical solutions or constructing realistic laboratory experiments. A carefully conducted laboratory hydraulic fracturing experiment has been published [32]. Figure 4shows our simulation of the experiment. The comparison is excellent not only in the created fracture geometry, but also the pumping parameters.

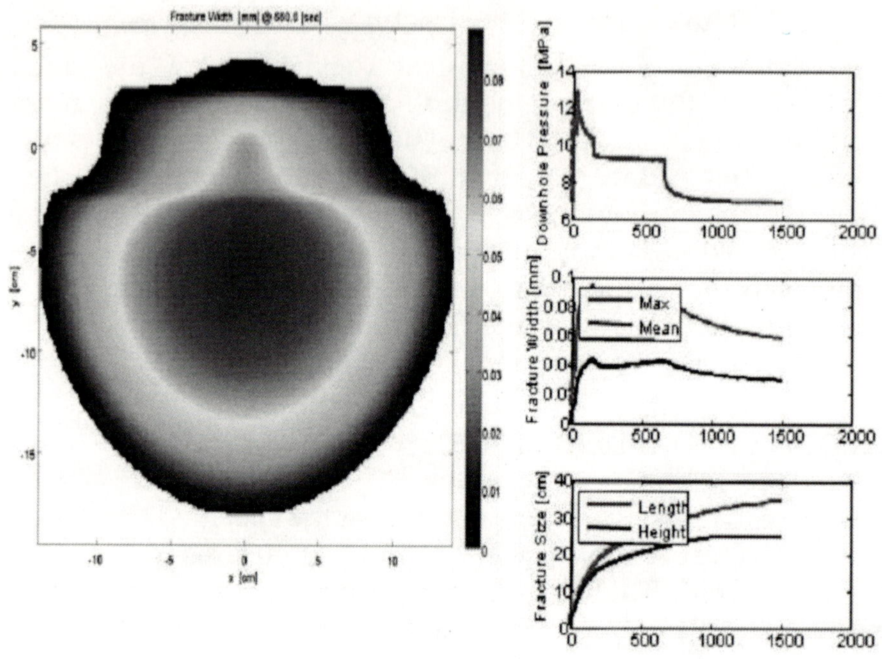

FIGURE 4. Comparing with experimental result of [32]

FRAC FLUID VISCOSITY AND FRACTURES HEIGHT GROWTH

A single stage of four hydraulic fractures is considered. Parameters chosen broadly resemble a typical shale gas development field case. The horizontal wellbore is placed at the relative depth of 0m as shown in Figure 5. Fracture injection points are spaced at 25m (82ft) apart and each fracture has 12 perforations. Fluid is injected into the wellbore at constant injection rate of 0.2 m^3/s (~75 bbl/min). Formation heterogeneities are limited to only in-situ stress variation. In this instance, the minimum horizontal stress gradient is set to 14kPa/m (0.62psi/ft) with no upper stress barrier to restrict height growth. The difference between minimum and maximum horizontal stress is chosen to be more than 20%, so that we do not cause the fracture direction to re-orient drastically. The key parameters are shown in Figure 5. The fluid leak-off factor is set to 1 with no spurt loss.

We consider two cases of distinctly different frac fluids, one with low viscosity (slick water) and the other a very high viscosity (gel frac). This corresponds to setting n = 0.8 and K = 0.05 for the low viscosity fluid case, and n = 0.8 and K = 0.5 for the high viscosity fluid in the power-law fluid model of Equation 7.

The injection parameters and the calculated total fracture volumes are shown in Figure 6. The simulated results are shown in Figure 7. As expected, the low viscosity fluid creates larger fractures with more fracture surface area at the expense of fracture aperture/width. Significantly, the low viscosity fluid creates a much larger fracture height. It is now interesting to see how the multiple fractures evolve in length and height as they compete in the presence of stress perturbation. Figure 8shows fracture development at the end of the pumping of high viscosity fluid. The outer two fractures grew slightly outward and are longer than the inner fractures. This is due to stress shadowing. Interestingly, the inner two fractures developed more height, and this behaviour will become even more pronounced when we look at low viscosity fluid.

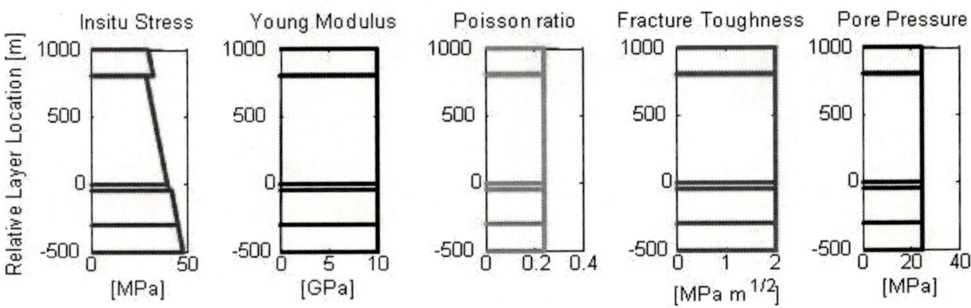

FIGURE 5. Formation input parameters for the case studies

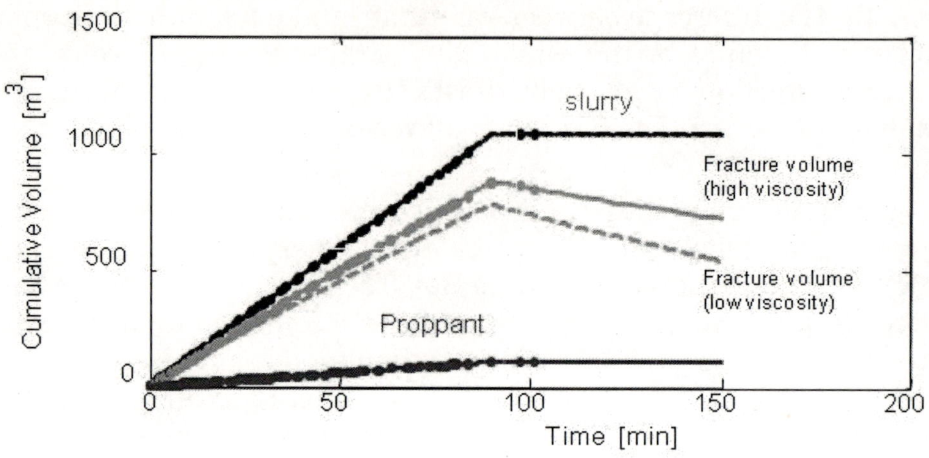

FIGURE 6. Injection volume and calculated fracture volumes

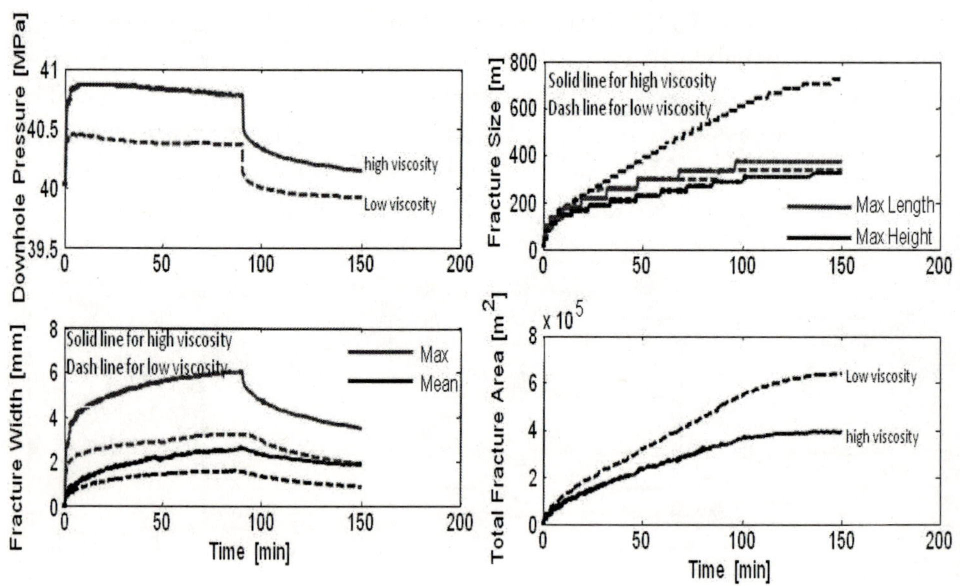

FIGURE 7. Calculated fracture parameters

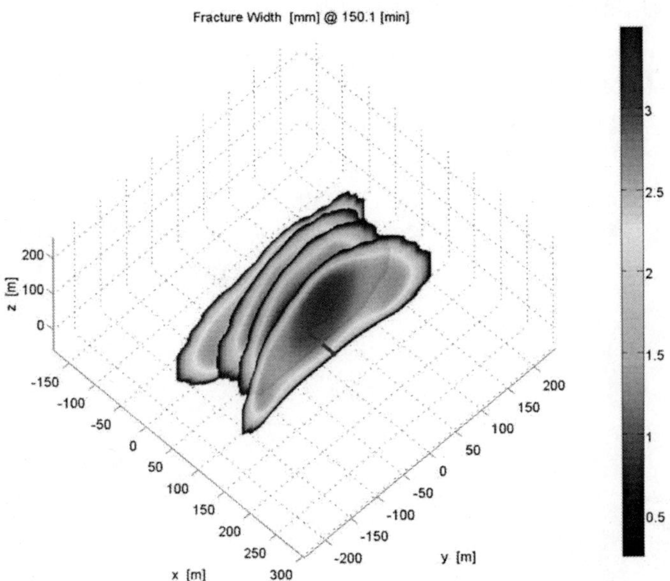

FIGURE 8. Fracture geometry with high viscosity frac fluid at 150 mins

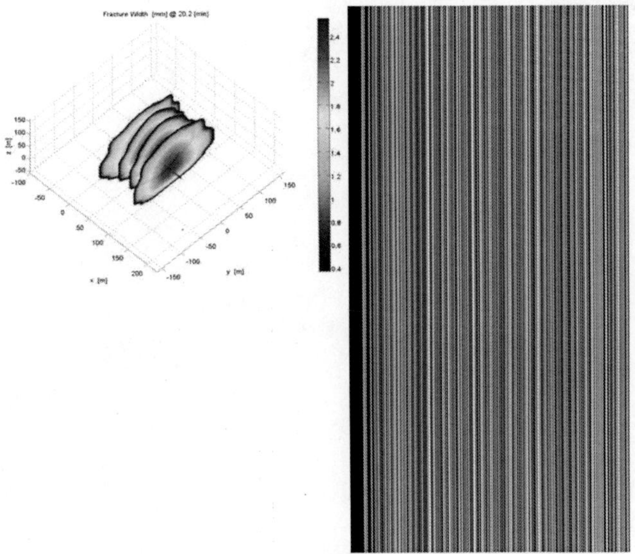

FIGURE 9. Fracture geometry of low viscosity frac fluid at 150 mins

Figure 9 shows the final fracture geometry of pumping low viscosity fluid (note that the scale in Figure 9 is different from that in Figure 8). Perhaps against intuition, the inner fractures grew appreciably in height. To understand this, we look at the evolution of fracture length and height growth over time as depicted in Figures 10 and 11. It becomes clear that as the inner fractures are constrained from growing in length because of competition with the outer fractures, they found relative freedom to grow upward instead.

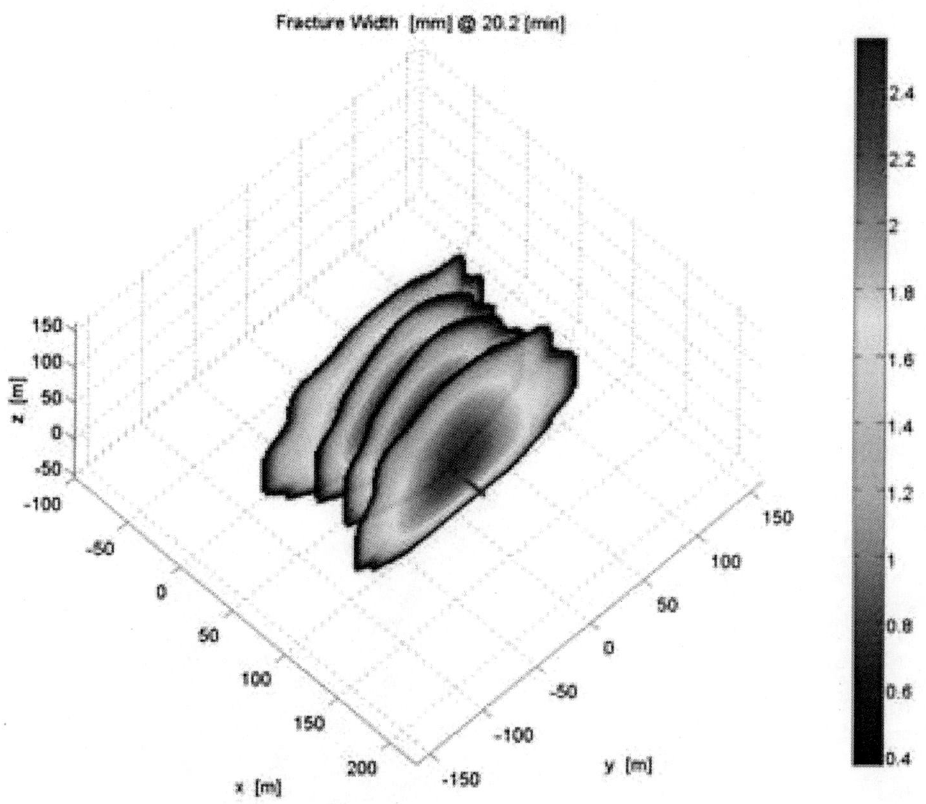

FIGURE 10. Fracture geometry of low viscosity frac fluid after 20 mins

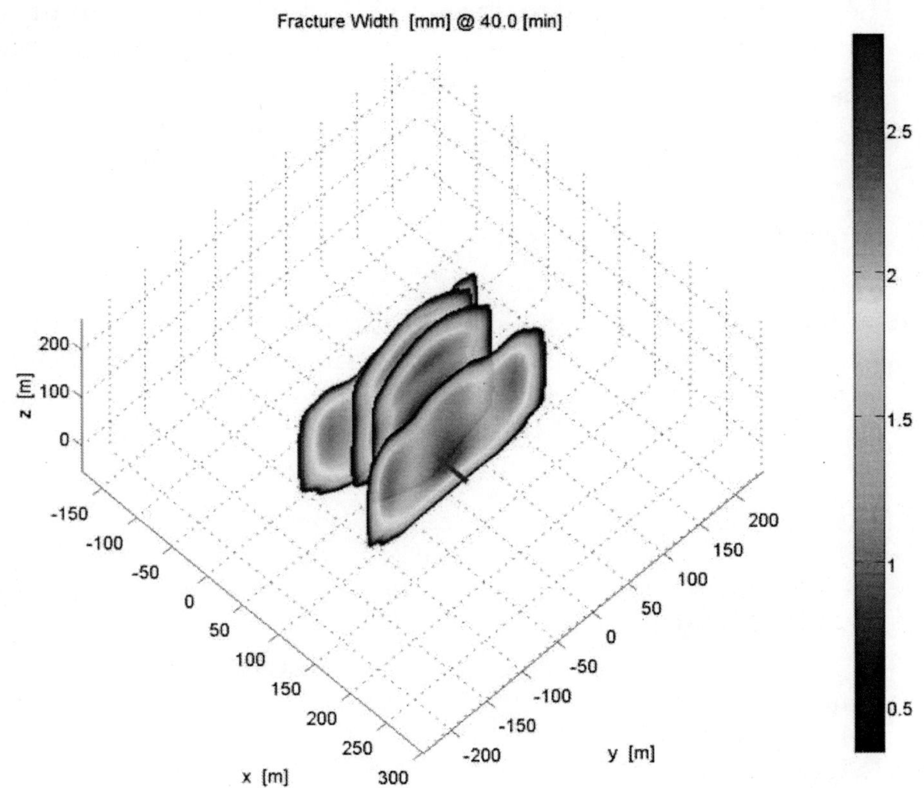

FIGURE 11. Fracture geometry of low viscosity frac fluid after 40 mins

DISCUSSION

At present, multiple fracture stimulations in horizontal wells often assume a uniform growth of all fractures, with the total volume of frac fluid more or less evenly distributed amongst the fractures. We have pointed out some of the potential impacts of stress shadows, which can affect treatment cost and production. Crucially, it has consequences in the understanding of subsurface development. For instance, if, instead of stimulating four fractures uniformly, only two or three of the fractures grow disproportionally, then some of the fractures may be unknowingly over-stimulated, resulting in excessive length or height growth. Normally this does not pose any major issue unless there are near-by wells or geological faults.

The non-planar 3D simulator presented here is capable of capturing the influence of key parameters such as injection rate, fracture spacing, formation properties, etc. on fracture design. Attempts are being made to extend the present numerical framework to include the interaction of hydraulic fractures with natural fractures.

ACKNOWLEDGEMENTS

The authors recognize the valuable input from Alexei Savitski, Anastasia Dobroskok, Mauricio Farinas, Ernesto Fonseca, Kaiming Xia, and Zongyu Zhai. Permission by Shell International Exploration and Production Inc. to publish is gratefully acknowledged.

REFERENCES

1. United States Department of Energy2009Modern shale gas development in the United States: A Primer.
2. P. N Mutalik, and B Gibson, 2008Case History of Sequential and Simultaneous Fracturing of the Barnett Shale in Parker County. Paper SPE 116124 presented at the SPE Annual Technical Conference and Exhibition, Denver, 2124September. DoiMS.
3. G Waters, B Dean, R Downie, K Kerrihard, L Austbo, B Mcpherson, 2009Simultaneous Fracturing of Adjacent Horizontal Wells in the Woodford Shale. Paper SPE119635 presented at the SPE Hydraulic Fracturing Technical Conference, Woodlands, TX, USA, 1921January. Doi:MS.
4. J Sierra, J Kaura, D Gualtieri, G Glasbergen, D Sarkar, D Johnson, 2008DTS Monitoring Data of Hydraulic Fracturing: Experiences and Lessons Learned. Paper SPE 116182 presented at the SPE Annual Technical Conference and Exhibition in Denver, Colorado, USA. September.
5. M. M Molenaar, D. J Hill, P Webster, E Fidan, B Birch, 2012First Downhole Application of Distributed Acoustic Sensing for Hydraulic-Fracturing Monitoring and Diagnostics. SPE Drilling & Completion, March, 3238
6. M. M Molenaar, E Fidan, and D. J Hill, 2012Real-Time Downhole Monitoring of Hydraulic Fracturing Treatments Using Fibre Optic Distributed Temperature and Acoustic Sensing. Paper SPE 152981 presented at the SPE/EAGE European Unconventional Resources

Conference and Exhibition in Vienna, Austria, March.

7. S. W Wong, M Geilikman, and G Xu, 2013Interaction of Multiple Hydraulic Fractures in Horizontal Wells. Paper SPE163982 presented at the Middle East Unconventional Gas Conference and Exhibition in Muscat, Oman. January.
8. J. E Olson, and K Wu, Sequential versus Simultaneous Multi-zone Fracturing in Horizontal Wells: Insights from a Non-planar, Multi-frac Numerical Model, Paper SPE 152602 presented at the SPE Hydraulic Fracturing technology Conference held in the Woodlands, Texas, February, 2012
9. J. C Morrill, and J. L Miskimins, 2012Optimizing Hydraulic Fracture Spacing in Unconventional Shales. Paper SPE 152595 presented at the SPE Hydraulic Fracturing Technology Conference in The Woodlands, Texas, USA. February.
10. M Rafiee, M. Y Soliman, and E Pirayesh, 2012Geomechanical Considerations in Hydraulic Fracturing Designs. Paper SPE 162637 presented at the SPE Canadian Unconventional Resources Conference in Calgary, Alberta, Canada. October.
11. N. P Roussel, and M. M Sharma, 2011Optimising Fracture Spacing and Sequencing in Horizontal-Well Fracturing. SPE Production and Operations. May, 173184
12. C. L Cipolla, E. P Lolon, M. J Mayerhofer, and N. R Warpinski, 2009Fracture Design Considerations in Horizontal Wells Drilled in Unconventional Gas Reservoirs. Paper SPE 119366 Presented at the SPE Hydraulic Fracturing Technology Conference, The Woodlands, Texas, USA, 1921January. DoiMS.
13. Y Cheng, 2009Boundary Element Analysis of the Stress Distribution Around Multiple Fractures: Implications for the Spacing of Perforation Clusters of Hydraulically Fractured Horizontal Wells. Paper SPE 125769 presented at the SPE Eastern Regional Meeting, Charleston, West Virginia, USA. September.
14. R Manchanda, and M. M Sharma, 2012Impact of Completion Design on Fracture Complexity in Horizontal Wells. Paper SPE 159899 presented at the SPE Annual Technical Conference and Exhibition in San Antonio, Texas, USA. October.
15. A. P Bunger, X Zhang, and R. G Jeffrey, 2012Parameters Affecting the Interaction Among Closely Spaced Hydraulic Fractures. SPE Journal. March, 292306
16. J. E Olson, 2008Multi-fracture Propagation Modeling: Applications to Hydraulic Fracturing in Shales and Tight Gas Sands. Paper ARMA 08327presented at the 42nd US Rock Mechanics Symposium in San

Francisco, California, June, 2008.

17. R Wu, O Kresse, X Weng, C Cohen, and H Gu, 2012Modeling of Interaction of Hydraulic Fractures in Complex Fracture Networks. Paper SPE 152052 presented at the Hydraulic Fracturing Technology Conference in The Woodlands, Texas, USA. February.
18. B Meyer, and L. W Bazan, 2011A Discrete Fracture Network Model for Hydraulically Induced Fractures- Theory, Parametric and Case Studies. SPE 140514. SPE Hydraulic Fracturing Technology Conference, 2426January 2011, The Woodlands, Texas, USA
19. G Xu, and S. W Wong, Interaction of Multiple Non-Planar Hydraulic Fractures in Horizontal Wells, Paper IPTC-17043 presented at the International petroleum technology Conference in Beijing, China, March 2013
20. J. C Jaeger, N. G. W Cook, and R. W Zimmerman, 2007Fundamentals of Rock Mechanics. Blackwell Publishing, 4th edition.
21. G. I Barenblatt, V. M Entov, and V. M Ryzhik, 1984Theory of Fluid Flows through Natural Rocks, Kluwer Academic Publishers.
22. I. N Sneddon, and M Lowengrub, 1969Crack Problems in the Classical Theory of Elasticity, John Willey & Sons, Inc.
23. N. R Warpinski, and L. W Teufel, 1987Influence of Geologic Discontinuities on Hydraulic Fracture Propagation, Journal of Petroleum Technology, February.
24. J Adachi, E Siebrits, A Peirce, and J Desroches, 2007Computer simulation of hydraulic fractures, Int. J. of Rock Mech. Mining Sci. 44739757
25. R. J Clifton, 1989Three Dimensional Fracture-Propagation models, in: Recent Advances in Hydraulic Fracturing, Editors: Gidley, J.L., S. A. Holditch, D.E. Nierode, and R.W. Veatch, Mono- graph Series, SPE.
26. E Siebrits, and A. P Peirce, 2002An efficient multi-layer planar 3D fracture growth algorithm using a fixed mesh approach, Int. J. Numer. Meth. Eng., 53691717
27. S. N Shah, 1993Rheological Characterization of Hydraulic Fracturing Slurries. SPE Production and Facilities. May, 123130
28. P Valko, and M. J Economides, 1995Hydraulic Fracturing Mechanics, Wiley, Chichester.
29. J Desroches, E Detournay, B Lenoach, P Papanastasiou, J. R. A Pearson, M Thiercelin, and A. H-D Cheng, 1994The crack tip region in hydraulic fracturing, Proc. R. Soc. London, Ser. A, 447, 39-48 (1994).
30. G Xu, 2000A Variational Boundary Integral Method for the Analysis of Three-Dimensional Cracks of Arbitrary Geometry in Anisotropic

Elastic Solids., J. Appl. Mech., 67403408

31. G Xu, and M Ortiz, 1993A Variational Boundary Integral Method for the Analysis of 3-D Cracks of Arbitrary Geometry Modeled as Continuous Distributions of Dislocation Loops, International Journal of Numerical Methods of Engineering, 3636753702
32. R Wu, A. P Bunger, R. G Jeffrey, and E Siebrits, 2008A Comparison of Numerical and Experimental Results of Hydraulic Fracture Growth into a Zone of Lower Confining Stress. Paper ARMA 08267presented at the 42nd US Rock Mechanics Symposium and 2nd U.S.-Canada Rock Mechanics Symposium, San Francisco, California, USA. June.

Chapter 9

COMPARISON OF HYDRAULIC AND CONVENTIONAL TENSILE STRENGTH TESTS

Michael Molenda[1], Ferdinand Stöckhert[1], Sebastian Brenne[1] and Michael Alber[1]

[1]Ruhr-University Bochum, Germany

Tensile strength is paramount for reliable simulation of hydraulic fracturing experiments on all scales. Tensile strength values depend strongly on the test method. Three different laboratory tests for tensile strength of rocks are compared. Test methods employed are the Brazilian disc test (BDT), modified tension test (MTT) and hydraulic fracturing experiments with hollow cylinders (MF = Mini Frac). Lithologies tested are a micritic limestone, a coarse-grained marble, a fine-grained Ruhrsandstone, a medium-grained rhyolite, a medium-/coarse-grained andesite and a medium grained sandstone. Test results reveal a relationship between the area under tensile stress at failure and the measured tensile strength. This relationship becomes visible when the area under tensile strength ranges over one order of magnitude from 450 to 4624 mm2. This observation becomes relevant when selecting the tensile strength values of lithologies.

Keywords: hydraulic fracturing, Brazilian Disc test, Modified Tension Test, Acoustic Emission, numerical simulation

INTRODUCTION

Tensile strength tests are widely applied in rock mechanics to obtain input parameters for planning of hydraulic fracturing on all scales. In literature only few experimental data sets are published dealing with samples size effects on tensile strength tests [1,2] or the comparison of different tensile tests in general [1,3]. Usually, results of laboratory tensile tests are taken to be size independent when used as input parameter for numerical studies at different spatial sizes.

We compare the results of 3 different, easily applicable laboratory tests for tensile strength of rocks. The sample set comprises a micritic limestone, a coarse-grained marble, a fine-grained Ruhr-Sandstone, a medium-grained rhyolite, a medium- /coarse-grained andesite and a medium grained sandstone. All tested rocks were characterized petrographically as well as by ultrasonic velocities, density, porosity, permeability, static, dynamic elastic moduli and compressive strength.

In order to determine the effects of specimen size on test results, we carried out BDT according to ISRM [4] with disc diameters of 30, 40, 50, 62, 75 and 84 mm, respectively. The recently presented MTT [5] was used as a tensile strength test with an approximately uniform tensile stress distribution. Hydraulic tensile strength was evaluated by MF experiments (core diameter 40 and 62 mm; borehole/diameter ratio 1:10) under uniaxial compression [6]. MF pressurization was performed with a constant fluid volume rate of 0.1 ml/s representing a stress rate of 0.3 MPa/s. In all tests relevant acoustic emission (AE) values have been evaluated to get additional information on the failure processes.

MATERIALS AND METHODS

Sample Material

To investigate the influence of rock properties on tensile test methods, six different rock types were tested. Bebertal sandstone, a medium grained Permian sandstone from a quarry near Magdeburg, Germany. Ruhrsandstone, a fine-grained and massive Carboniferous arcose from the Ruhr area in Germany. A medium to coarse grained, jointed Permian andesite from the Doenstedt Eiche quarry near Doenstedt, Germany. A medium grained, highly jointed Permian rhyolite from the Holzmuehlental quarry near Flechtingen, Germany. A micritic Jurassic limestone from a quarry near Treuchtlingen, Germany and a coarse grained marble from Carrara, Italy. The rocks' petrophysical properties, namely bulk density, grain density, compressional wave speed, porosity, permeability, cohesion and friction angle are listed inTable 1.

PETROPHYSICAL CHARACTERIZATION

Dry densities are calculated geometrically based on geometrical properties, grain densities are measured according to DIN 18124. Compressional wave velocities are measured at each core with a Geotron USG 40/UST 50-12 at room temperature and in dry condition. Porosities are derived from the difference between grain density and geometrical density of the oven-dried samples. Permeabilities are evaluated via a constant head test on the hollow cylinder samples used for the MF tests [7]. Bebertal-sandstones are permeable enough to use a simple axial flow-through test with a maximum pressure difference of up to 3 bars. The samples are sealed off with rubber jackets to minimize water-flow along the sample surface. Unconfined compressive strengths and static moduli of elasticity are measured by uniaxial compressive tests [8].

TABLE 1. Averaged values of petrophysical properties of the rock samples. ρ_d dry bulk density, ρ_s grain density, vp compressional wave velocity, Φ porosity, k permeability, c cohesion, φ friction angle.

Rock type (location)	ρ_d [g/cm³]	ρ_s [g/cm³]	vp [m/s]	Φ [%]	k [m²]	c [MPa]	φ [°]
Marble (Carrara)	2.71 ±0.002	2.721 ±0.003	5.67 ±0.06	0.40	1E-19	29	22
Limestone (Treuchtlingen)	2.56 ±0.008	2.713 ±0.002	5.59 ±0.05	5.64	1E-18	27	53
Ruhrsandstone (Ruhr area)	2.57 ±0.006	2.688 ±0.008	4.61 ±0.13	4.39	8E-18	36	50
Rhyolite (Flechtingen)	2.63 ±0.015	2.657 ±0.011	5.39 ±0.34	1.02	9E-19	20-36	55
Andesite (Dönstedt)	2.72 ±0.023	2.734 ±0.006	5.26 ±0.28	0.51	-	20-41	50
Sandstone (Bebertal)	2.66 ±0.061	2.44 ±0.059	3.61 ±0.61	8.27	11E-15	15	45

TESTING PROCEDURE OF THE TENSILE STRENGTH TESTS

All experiments are performed in a stiff servo-hydraulic loading frame from Material Testing Systems (MTS) with a load capacity of 4000 kN. For further details on the technical specifications see Table 2.

TABLE 2. Technical specifications of the measurement system.

Device (manufacturer) name	**max. capacity**	**accuracy**	**BDT**	**MTT**	**MF**
Axial load cell (Althen) CPA-50	500 KN	± 100 N	x	x	x
Axial displ. transducer (Scheavitz) MHR 250 LVDT 1 & 2	6.3 mm	$\pm 1{\cdot}10^{-4}$ mm	x	x	x
Displ. transducer at pressure intensifier (HBM) WA 100 mm LVDT 3	100 mm	$\pm 1{\cdot}10^{-3}$ mm			x
Load cell for Hoek Cell					
Load cell for pressure intensifier (Burster) 8219R-3000	300 MPa	± 0,03 MPa			x

Acoustic Emission (AE) signals are acquired with an AMSY-5 Acoustic Emission Measurement System (Vallen Systeme GmbH, Germany) equipped with up to 6 Sensors of type VS150-M. The Sensors are sensitive in a frequency range of 100-450 kHz with a resonance frequency of 150 kHz and a preamplification of 34 dB_{AE}. Due to machine noise in the range below 100 kHz incoming signals are filtered by a digital bandpass-filter in a frequency range of 95-850 kHz. AE data are sampled by a sampling rate of 10 kHz. The sensors are fixed using hot-melt adhesive to ensure best coupling characteristics. Pencil-break tests (Hsu-Nielsen source) and sensor pulsing runs (active acoustic emission by one sensor) are used to ensure good sensor coupling of the sensor on the sample.

HYDRAULIC FRACTURING CORE EXPERIMENTS (MF) PROCEDURE

Minifrac experiments are carried out mainly on 40 mm cores with a borehole diameter of 4mm. Furthermore some 62 mm cores with a borehole of 6 mm diameter are tested. The samples are loaded axially up to 5 MPa to ensure that the packer mechanism is tight and seals off the borehole openings at the top and at the bottom. The borehole pressure was raised servo controlled with a fixed volume rate of 0.1 ml/s that results in a pressure rate of approximately 0.3 MPa/s. All MF tests are monitored by Acoustic Emissions with four sensors glued directly to the samples and a fifth sensor placed at the incoming hydraulic line.

BRAZILIAN DISC TESTS (BDT) PROCEDURE

All Brazilian disc tests are carried out following the ISRM suggested method [4] at a load rate of 200 N/s. Disc diameters used are 30, 40, 50, 62, 75 and 84 mm, whereas the length to diameter ratio (L/D) was constant at 0.5. All tests are monitored by one AE-sensor glued directly in the middle of the disc specimen. The size dependency is tested with discs from Ruhrsandstone, marble, rhyolite and limestone.

MODIFIED TENSION TEST (MTT) PROCEDURE

The MTT tests are driven load controlled at a rate of 200 N/s that corresponds to a stress rate of 0.02 MPa/s. The axial force is applied from the top (Figure 1). MTT test samples are observed by up to 6 AE-Sensors glued directly to the specimen. The samples were overcored with 62 mm and 30 mm diameters where the overlapping height is 1/3 of the total sample height (Figure 1). The centralizing of the drills was achieved by using a former plate to adjust the sample before drilling. Despite assiduously arrangement the eccentricity of the overcoring was in the range of up to 3 mm due to the imprecise vertical guidance of a standard drilling machine. In order to test the influence of eccentricity we also prepared samples with an eccentricity of 14 mm.

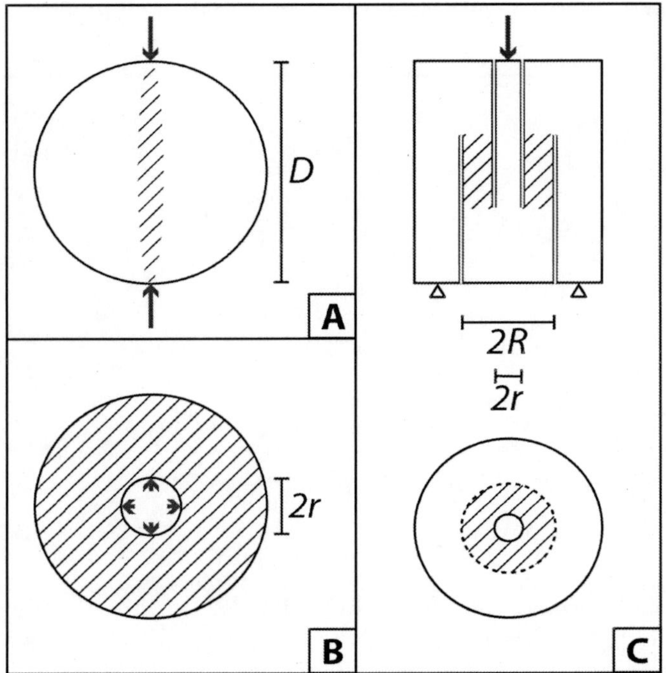

FIGURE 1. Sketches of the three tensile test methods. A: BDT side view, B: MF top view, C: MTT side view cross section (upper) and top view (lower).

EXPERIMENTAL RESULTS

BRAZILIAN DISC TEST SIZE DEPENDENCY

The size dependency of the absolute size of the Brazilian disc test discs on the tensile strength is shown in Figure 2. Overall 138 Brazilian disc tests are undertaken for up to 6 sizes and four lithologies. The disc diameters, ranging from 30-84 mm, represent the sizes that are mostly tested in laboratories to determine the BDT tensile strength of rock samples. The results of the size dependency tests show no significant relationship between the sizes of the tested disc to its calculated tensile strength as long as the length to diameter ratio is held constant as suggested by the ISRM suggested method at a value of 0.5 [4]. There is a marginal tendency for the standard deviation of the tensile strength to decrease with increasing disc size.

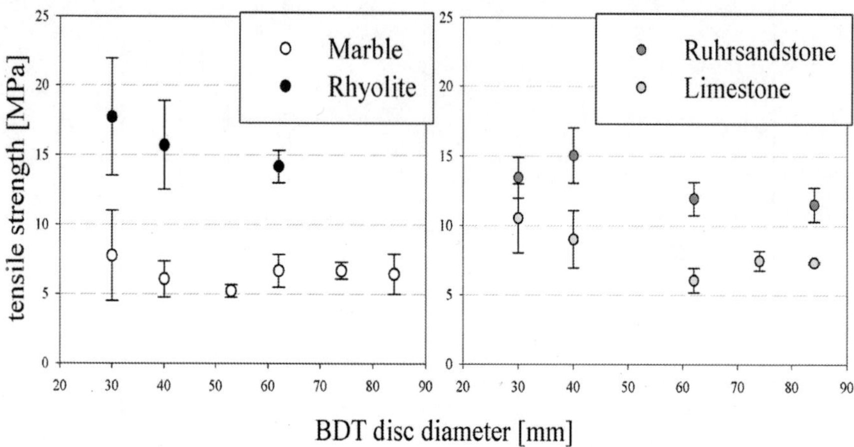

FIGURE 2. Size dependency of the BDT disc size on the tensile strength for four lithologies. Circles represent the mean values, bars stand for the standard deviation.

MF, BDT AND MTT TENSILE STRENGTH RESULTS

Three different methods for the determination of tensile strength are compared regarding their results. 201 Brazilian disc tests, 31 Minifrac tests and 15 Modified tension tests form the basis of the data evaluation, where σBDTt, σMFt and σMTTt are the tensile strengths indexed by the used method. BDT tensile strengths are calculated as follows [4].

$$\text{BDT: } \sigma BDTt = 2P/\pi Dt$$

Where P is the force at failure, D is the disc diameter and t the disc thickness.

For the MF tests, assuming the rocks to be nearly impermeable and therefore neglecting a relevant pore pressure influence the tensile strength is given directly by the breakdown pressure Pb [9].

$$\text{MF: } \sigma MFt = Pb$$

MTT tensile strengths are evaluated by the formula given by [5].

$$MTT: \sigma MTTt = Fmax/ATZ = Fmax/(R2\pi - r2\pi)$$

Where R2 and r2 are the outer and inner radius, respectively (Figure 1). Mean values, standard deviations and total number of tests for all three testmethods can be found in Table 3.

TABLE 3. Comparison of tensile strength out of three test methods.

Lithology	Test method	Mean [MPa]	Std. dev. [MPa]	N [-]
Ruhrsandstone	BDT	13.2	2.1	32
	MF	19.0	3.0	10
	MTT	5.8	1.0	3
rhyolite	BDT	15.8	3.2	39
	MF	20.1	5.5	5
	MTT	4.9	1.4	2
limestone	BDT	8.2	2.2	36
	MF	10.2	1.7	5
	MTT	4.8	1.0	2
marble	BDT	6.4	1.5	32
	MF	7.8	1.3	4
	MTT	4.3	1.2	2
andesite	BDT	14.6	4.5	23
	MF	14.4	5.1	4
	MTT	8.7	4.4	3

Bebertal sandstone	BDT	4.1	1.2	39
	MF	4.3	2.0	3
	MTT MTT eccentric	2.4 1.0	- 5E-3	1 2

One of the main observations is the very low tensile strength measured with the Modified Tension Test method. The MTT results mean values are in the range of 66 % down to 31 % of those obtained with the BDT. In addition to the low tensile strengths obtained by the MTT an eccentricity of the overcoring yields to an additional underestimation of the tensile strength values. The BDT and MF results seem to be more similar. The BDT results lie in the range of 70 % to 100 % of the MF tensile strength, so the MF test yields the highest tensile strengths and also to the highest standard deviations. All measurements are visualized in Figure 3. Doenstedt andesite and Flechtingen rhyolite tensile strengths have the highest standard deviations of the tested rock types. This variance is due to the high amount of natural joints that are assumed to have a different tensile strength with respect to the intact parts. Therefore the tensile strength scattering is the result of the material heterogeneity itself.

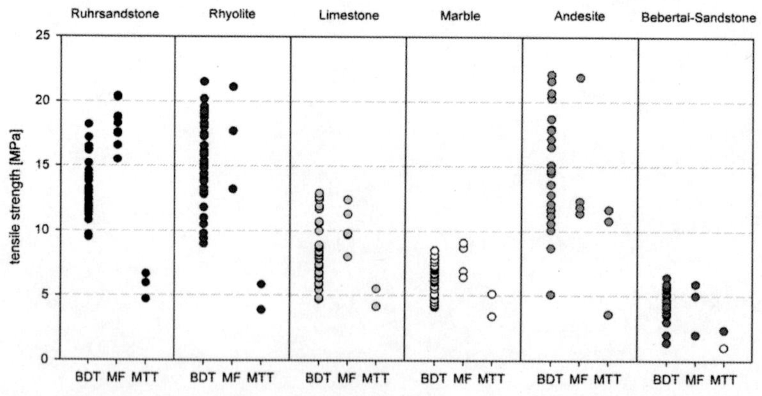

FIGURE 3. Results of all tensile strength test. BDT: Brazilian Disc Test, MF: Minifrac, MTT: Modified Tension Test. Hollow circle in Bebertal sandstone MTT tests represents two results of the highly eccentrical MTT tests.

ACOUSTIC EMISSIONS RESULTS

Acoustic emission data obtained during the tests give rough insights into the failure processes. It is obvious that all tests end with a spalling of the specimens in parts due to a complete tensile failure. Simple AE count analysis show that the BDT is accompanied with an immense hit-rate long before total failure in comparison to the relatively quiet pre-failure phases of the MF and MTT tests. In good agreement with theoretical considerations of the stress distribution in the Brazil disc [1] these events are most likely due to compressional failure at the top and bottom of the disc, accompanied with crack propagation and coalescence before peak load (Figure 4).

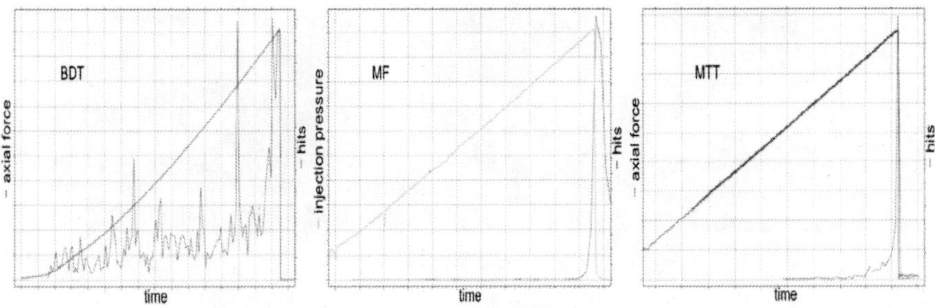

FIGURE 4. AE hits per 0.5 sec., BDT left, MF middle and MTT right showing the huge difference in AE hits before total failure of the sample.

NUMERICAL MODEL

We investigate the effect of eccentricity of the overcoring for the MTT samples by a numerical simulation. A finite element study that has been performed by Plinninger et al. [10] that shows a uniform tensile stress distribution in the annulus of the test samples. It is arguably if this model is the right tool for modeling a tensile stress distribution in rock samples prior to failure. A simple linear elastic 3D FEM model reveals tensile stress concentrations at the edges of the rims in the sample (Figure 5). Fractures may be initiate there at relative low axial forces.

During preparation of the samples it becomes obvious that exact centralization of the inner overcoring is not always given. Two Bebertal sandstone samples were prepared with a eccentricity of 14 mm resulting in a minimum rim width of 2 mm instead of 16 mm for a perfectly centralized sample. The average eccentricity of our samples is in the range of up to 3 mm. Tensile stress redistribution due to eccentricity is modeled as well and can easily double the tensile stress in the thinner rim of the annulus (Figure 5).

FIGURE 5. Slice through a linear elastic 3D FEM model of MTT tensile test. Values of axial stress are given in MPa where negative values stand for tensile stress. Left model represents a perfectly centralized sample. Right model shows the stress distribution for a eccentricity of 6 mm towards the left edge.

DISCUSSION

247 tensile strength test results of BDT, MF and MTT tests vary considerable within one lithology (Figure 6). Therefore it is not trivial to give a reliable prediction of the tensile strength parameter. Results of the BDT tests show no significant variation with respect to the specimen size, as long as the aspect ratio is held constant. Nevertheless the tensile strength data scattering is high, so that it may obscure existing trends. Acoustic Emission evaluation shows that during the BDT multiple fracturing mechanisms are present. Before total fracturing of the sample by a tensile rupture there is a high amount of AE activity that is most likely related to

compressional failure at the top and bottom of the disc. Beside this, compressional stress concentrations and the inhomogeneous tensile stress distribution may lead to tensile cracks before peak load. MF results lead to the highest tensile strengths in this comparison where there seem to be no differences in tensile strength when using a 4 mm or a 6 m borehole for pressurization. Again one has to take into account that the high amount of tensile strength scattering for these tests inhibits a statement regarding a borehole size dependency.

The results of the MTT tests give the lowest tensile strengths and very low standard deviations. Latter may be related to the small amount of testes MTT per lithology. Furthermore all MTT are prepared using the same sample sizes. A major problem of the MTT experiments is the centralization of the boreholes. An eccentricity yields to a significant inhomogeneity of the tensile stress distribution in the sample (Figure 5). Numerical simulations of the MTT eccentricity effect together with the two eccentric MTT samples (Figure 3) show that the calculated tensile strength may be underestimated massively. One reason for the apparently lower tensile strength measured using the MMT might be the applicability ofEquation (3). In deriving the equation, it was assumed that, when the peak load is approached, the tensile stress distribution is almost uniform in the area defined as ATZ [5]. This may only be true if the material is highly ductile. However, for brittle rocks, especially for highly fractured rocks, fracture propagation may occur and lead to ultimate failure at a much lower load as suggested by Equation (3) due to stress concentration (Figure 5).

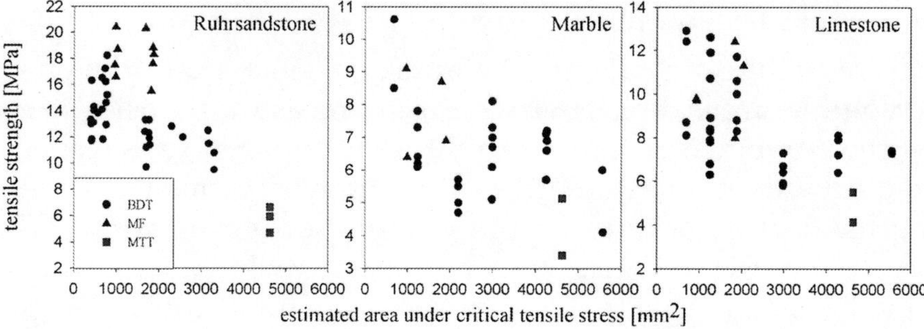

FIGURE 6. Tensile strength results plotted against the assumed area under

tension. BDT: diameter x thickness, MF: surface area of the borehole and MTT: twice the surface area between the outer and inner borehole, upper and lower.

Testmethod	BDT	MF	MTT
Area under 1,51,5	450-3400 m"2	1005-3393 m"2	4624 m"2
Calculation	$D \cdot t$	$2 \cdot \pi \cdot rbh \cdot l$	$2 \cdot (\pi \cdot (R2 - r2))$

TABLE 4. Estimated area subjected to tensile stress for the different tensile tests. D: BDT disc diameter, t: BDT disc thickness, rbh: MF borehole radius, l: MF sample height, R: MTT outer borehole radius, r: MTT inner borehole radius.

Main difference in all experiments and the reason for choosing these are the areas that are under tensile stress at the point of failure. The calculated tensile strengths compared to the area perpendicular to the maximum tensile stress show a negative trend for the tensile strength with increasing area being set under tensile stress. That is reasonable in terms of the statistical theory of strength. Especially for the igneous rocks it seems evident, that the probability to set a healed joint under a critical tension rises with the size of the sample volume that is under tensional stress. For the selection of the tensile strength test one should keep in mind that depending on the lithology the apparent tensile strength appears to be a function of the area, or more exact of the volume under tensile stress. Thus, for a relative homogeneous rock a less severe reduction of the measured tensile strength with size will be visible as it will be at the highly fractured igneous rocks tested in this study.

It is arguable and may not be appropriate to study the effect of area/volume under tensile stress on the measured tensile strength using the combined results from different types of tests, especially if the different tests tend to give different average measured tensile strengths. Furthermore the negative trend of tensile strength with respect to the stressed area/volume is not that obvious for the single test methods. Especially the assumption of uniform tensile stress distribution close to peak load in the annulus [5] for the MTT samples seems not to be comprehensible. It may hold for ductile materials but not for brittle ones. Therefore the validity of equation (3) for the calculation of the tensile strength is questionable. Nevertheless the resulting tensile strengths are treated as the same rock property when used as input parameters for calculations. This is very problematic due to its huge variation as shown in the tests. The correlation of the calculated tensile strength with the stressed area/volume is one possible approach to account for the decreasing apparent tensile strength behavior.

ACKNOWLEDGEMENTS

This work is funded by the Federal Ministry of Environment, Nature Conservation and Nuclear Safety (funding mark 0325279B). Special thank goes to Kirsten Bartmann and Sabrina Hoenig for laboratory work and data evaluation done during their Master Theses.

REFERENCES

1. M Mellor, I Hawkes, Measurement of tensile strength by diametral compression of discs and annuli. Engineering Geology. Elsevier; 197153173225
2. K Thuro, R. J Plinninger, S Zäh, S Schütz, Scale effects in rock strength properties. Part 1: Unconfined compressive test and Brazilian test. Rock Mechanics-A Challenge for Society.-881 p., Proceedings of the ISRM Regional Symposium Eurock. 200116974
3. J. A Hudson, E Brown, F Rummel, The controlled failure of rock discs and rings loaded in diametral compression. International Journal of Rock Mechanics and Mining Sciences & Geomechanics Abstracts. Elsevier; 1972922414IN1-IN4, 245-8.
4. Z. T Bieniawski, I Hawkes, Suggested Methods for Determining Tensile Strength of Rock Materials. International Journal of Rock Mechanics and Mining Sciences & Geomechanics. Abst, 19781599103
5. Thomée DIBWolski DGK, Plinninger RJ. The Modified Tension Test (MTT)-Evaluation and Testing Experiences with a New and Simple Direct Tension Test. Eurock 2004Oct 23.
6. D Schmitt, M Zoback, Infiltration effects in the tensile rupture of thin walled cylinders of glass and granite: Implications for the hydraulic fracturing breakdown equation. International Journal of Rock Mechanics and Mining Sciences & Geomechanics Abstracts. Elsevier; 1993303289303
7. Selvadurai APSJenner L. Radial Flow Permeability Testing of an

Argillaceous Limestone. Ground Water. 20135110007

8. C Fairhurst, J. A Hudson, Draft ISRM suggested method for the complete stress-strain curve for intact rock in uniaxial compression. Int J Rock Mech Min. Elsevier; 199936327989
9. M Zoback, F Rummel, R Jung, C Raleigh, Laboratory hydraulic fracturing experiments in intact and pre-fractured rock. International Journal of Rock Mechanics and Mining Sciences & Geomechanics Abstracts. Elsevier; 19771424958
10. R. J Plinninger, K Wolski, G Spaun, B Thomée, K Schikora, Experimental and model studies on the Modified Tension Test (MTT)-a new and simple testing method for direct tension tests. GeoTechnical Measurements and Modelling-Karlsruhe. 200320033616

Chapter 10

EFFECT OF FLOW RATE AND VISCOSITY ON COMPLEX FRACTURE DEVELOPMENT IN UFM MODEL

Olga Kresse[1], Xiaowei Weng[1], Dimitry Chuprakov[2], Romain Prioul[2] and Charles Cohen[1]

[1] Schlumberger, Sugar Land, USA

[2] Schlumberger Doll Research, Boston, USA

INTRODUCTION

It is believed that complexity of the fracture network created during hydraulic fracturing treatments in formations with pre-existing natural fractures is caused mostly by the interaction between hydraulic and natural fractures. The understanding and proper modelling of the mechanism of hydraulic-natural fractures interactions are keys to explain fracture complexity and the micro seismic events observed during hydraulic fracturing treatments, and therefore to properly predict production.

When a hydraulic fracture (HF) intercepts a natural fracture (NF) it can cross the NF, open (dilate) the NF, or be arrested at NF. If the hydraulic fracture crosses the natural fracture, it remains planar, with the possibility to open the intersected NF if the fluid pressure at the intersection exceeds the effective stress acting on the NF. If the HF does not cross the NF, it can dilate and eventually propagate into the NF, which leads to more complex fracture network. So the crossing criterion in general controls the complexity of the resulting fracture network.

The interaction between HF and NF depends on the in-situ rock stresses, mechanical properties of the rock, properties of natural fractures, and the hydraulic fracture treatment parameters including fracturing fluid properties and injection rate. During the last decades, extensive theoretical, numerical, and experimental work has been done to investigate, explain, and develop the rules controlling HF/NF interaction. Among the main contributions to this topic are the work listed in references [1-15].

Most of the existing crossing models do not take into account fluid properties due to the complexity of modelling fluid-solid interaction in the vicinity of the intersection, so crossing behavior is explained purely from elasticity point of view. Field and laboratory observations, however, show that fluid properties are important and should be accounted for [9, 16].

It is well known that the micro seismic events cloud is related to the hydraulic fracture propagation pattern which in turn strongly depends on the HF/NF interaction rules [17].

Figure 1 shows the micro seismic events observed in the same well first treated with a cross-linked gel, and then re-fractured with slick water [16]. Cross-linked gel was pumped at 70 bpm for about 3 hours with sand concentration ramped up to 3 ppg. Most of the micro seismic activity suggests longitudinal fracturing with only modest activation of natural fractures, resulting in a narrow stimulated network (less than 500 feet from the wellbore in many sections of the lateral), as seen in Figure 1a with resulting Stimulated Reservoir Volume (SRV) equal to 430 million ft^3. During the full re-frac conducted the following day 60,000 bbl of slick water and 285,000 lb of sand was pumped at 125-130 bpm for most of the treatment lasting 6.5 hours. The stimulated network was approximately 1500ft wide and 3,000 ft long (Figure 1b) with considerable height growth and SRV of 1450 million ft^3. Clearly, the re-fracturing treatment stimulated a much larger volume of rock than the initial gel treatment (1450 million ft^3 vs 430 million ft^3), and showed the patterns of development that suggested the opening of both northeast and northwest trending fractures [16].

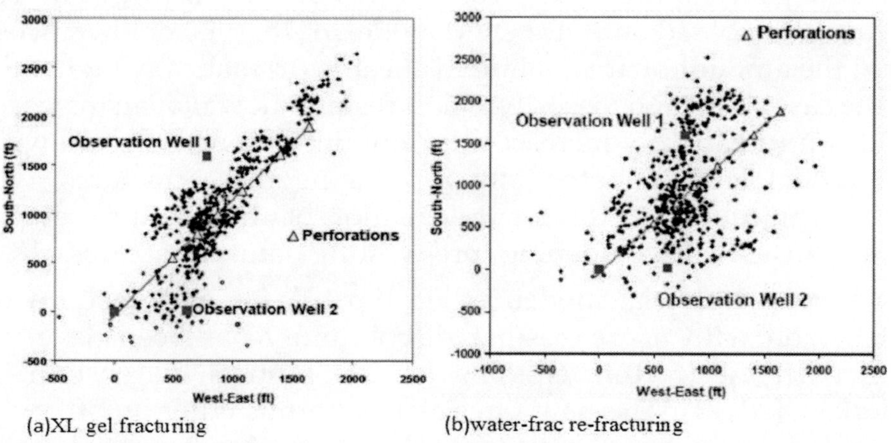

(a)XL gel fracturing (b)water-frac re-fracturing

FIGURE 1. Single-well micro seismic event locations for XL gel stimulation and water-frac re-fracturing treatment, horizontal Barnett Shale well [16]

This field example indicates the importance of proper consideration of fluid properties when modelling the interaction of hydraulic fractures with pre-existing natural fractures. In general it is observed that for the same field conditions more viscous fluid tends to cross the natural fractures more easily, while slick water tends to penetrate into the natural fractures more easily and open them without crossing. Pumping rate as well as rock properties should also be taken into account.

The importance of fluid properties on the created hydraulic fracture network has been mentioned in some experimental and numerical studies [9, 18, 19]. The experimental study of the influence of flow rate and fracturing fluid viscosity on the hydraulic fracture geometry have been performed in [9] based on analysis of different $Q\mu$ value (product of the injection rate and fracturing fluid viscosity). The experiments show that with low $Q\mu$ value fluid tends to leak into the pre-existing discontinuities despite the influence of fluid pressure and once the discontinuity accepts fluid, the pressure can rise far above the confining stress without inducing new fractures. With large $Q\mu$ value the hydraulic fracture tends to cross natural fracture due to increase of the pressurization rate.

The influence of fluid injection rate and viscosity on the amount of the tensile failure in the rock with natural fractures has been investigated based on 3DEC DEM model in [18, 19]. For low viscosity fluid the amount of area failing in shear is dramatically higher than in the case with high viscosity. Their results show that an increase in injection rate greatly increases the amount of tensile failure within the model leading potentially to creating more fractures, while a lower injection rate favors the creation of shear failure resulting mostly in activating (opening) pre-existing natural fractures.

A new analytical model, called OpenT, for hydraulic fracture interaction with a pre-existing discontinuity has been developed to predict the fracture crossing or deflection at the encountered interface [20, 21]. The new physically rigorous criterion of fracture re-initiation at the discontinuity has been implemented, which combines both stress criterion and energy release rate. It has been shown that the OpenT model adequately predicts the fracture crossing of non-cohesive frictional interfaces observed in various laboratory experiments with different interface orientations with respect to hydraulic fractures [21].

The new crossing model predicts the dimensions of open and sliding zones created at cohesive and non-cohesive interfaces after the intersection with a fluid-driven fracture. Such information can be valuable, for example, in passive micro seismic monitoring of fracture treatments in naturally fractured formations. By thoroughly examining the stress field generated by the hydraulic fracture and activated open and sliding zones at the discontinuity, it was shown that the new fracture initiation point is shifted along the inclined interface. The model predicts the offset of a secondary fracture as a function of the geometrical, loading, and mechanical parameters of the system, such as the fracture-interaction angle, in-situ stress components and fracture toughness in rock.

New OpenT crossing model incorporates the influence of rock properties (local horizontal stresses, rock tensile strength, toughness, pore pressure, Young's modulus, Poisson ratio), natural fracture properties (friction coefficient, toughness, cohesion, permeability), intersection angle between hydraulic and natural fractures, fracturing fluid properties (viscosity, tip pressure), and injection rate to define crossing rules.

This new OpenT model has been validated against laboratory experiments and against rigorous numerical models [3,20,21]. It was incorporated into the UFM model that simulates complex fracture network propagation in a formation with pre-existing natural fractures [22-24]. We present several UFM test cases showing the influence of injection rate and fluid viscosity on the generated hydraulic fracture footprint and production impact by comparing of two crossing criteria, the extended Renshaw & Pollard (hereafter referred as eRP) [14, 15] and the OpenT.

UFM model specifics

A complex fracture network model, referred to as Unconventional Fracture Model (UFM), had recently been developed [22,23,24]. The model simulates the fracture propagation, rock deformation, and fluid flow in the complex fracture network created during a fracture treatment. The model solves the fully coupled problem of fluid flow in the fracture network and the elastic deformation of the fractures, which has similar assumptions and governing equations as conventional pseudo-3D fracture models. Transport equations are solved for each component of the fluids and prop pants pumped. A key difference between UFM and the conventional planar fracture model is being able to simulate the interaction of hydraulic fractures with pre-existing natural fractures, i.e., determine whether a hydraulic fracture propagates through or is arrested by a natural fracture when they intersect and subsequently propagates along the natural fracture.

To properly simulate the propagation of multiple or complex fractures, the fracture model takes into account the interaction among adjacent hydraulic fracture branches, often referred to as "stress shadow" effect. It is well known that when a single planar hydraulic fracture is opened under a finite fluid net pressure, it exerts a stress field on the surrounding rock that is proportional to the net pressure. The details of stress shadow effect implemented in UFM are presented in [24].

The branching of the hydraulic fracture at the intersection with the natural fracture gives rise to the development of a complex fracture network. A crossing model that is extended from the Renshaw-Pollard [10] interface crossing criterion, applicable to any intersection angle, has been developed [14], validated against the experimental data [15], and was integrated in the UFM. The previous crossing model, showing good comparison with existing experimental data, does not account for the fluid impact on the crossing pattern.

The new crossing model (OpenT) which accounts for the fluid properties is presented in short below and is implemented in a new version of UFM.

New crossing model in UFM

There are a few analytical criteria describing the mechanical HF-NF interaction developed in the past [10, 11, 13, 14]. With their relative simplicity they do not take into account the influence of the fluid injection into the hydraulic fracture and the fluid infiltration into the natural fracture after contact. These criteria were designed to capture the effect of the fracture approach angle, the NF friction coefficient and the anisotropy of the in-situ stresses. To improve the description of HF-NF interaction a new analytical model that takes the mechanical influence of the HF opening and the hydraulic permeability of the NF into account has been developed.

The analytical model of the HF-NF interaction (OpenT) solves the problem of the elastic perturbation of the NF at the contact with the blunted HF tip, which is represented by a uniformly open slot (i.e. giving its name OpenT) [21]. The opening of the HF at the junction point w_T (blunted tip) develops soon after contact, and approaches the value of the average opening of the hydraulic fracture

$\overline{w}$, defined by the injection rate Q and the fluid viscosity μ. In a viscosity-dominated regime, the average opening of the KGD fracture with half-length L and height H can be estimated as [25]

$$\overline{w} = 2.53\left[\frac{Q\mu L^2}{E'H}\right]^{1/4} \quad (1)$$

where

$$E'=E/(1-v^2)$$

, E is the Young modulus, v is the Poisson coefficient. The OpenT model looks for the solution of the elastic problem for the NF perturbed by the HF, and outputs the profiles and boundaries of the opening and sliding zones as a result of the contact (bo and bs respectively shown in Figure 2, left).

The solution shows that the spatial extent of the open and sliding zones strongly depends on the fluid pressure inside the activated part of the NF. The larger the inner fluid pressure, the larger the open and sliding zones at the NF are. Consequently, it is expected that after the HF-NF contact, the injected fracturing fluid will gradually penetrate the NF with finite hydraulic permeability κ and thus enhance the inner fluid pressure within the NF, p_{NF}.

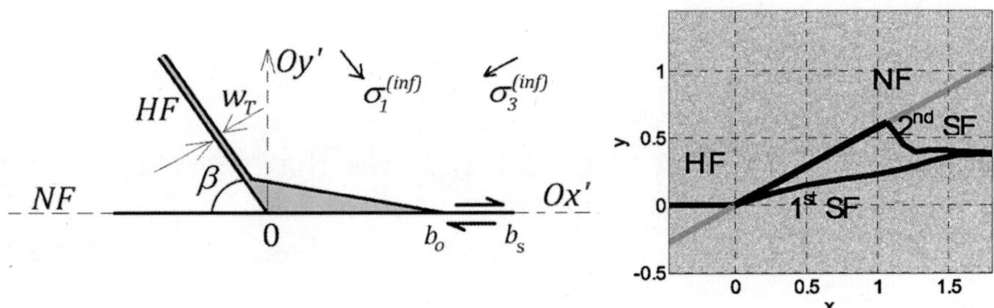

FIGURE 2. Left –Schematic diagram of the HF-NF interaction. Right – result of the computed HF/NF interaction with the initiation of two secondary fractures and their subsequent propagation.

The average pressure of the fracturing fluid penetrated the NF can be as approximated by the following function of the contact time *t*

$$p_{NF}(t) = p_f \tanh\left(\sqrt{\frac{2\kappa p_f}{\mu b_s^2} t}\right) \tag{2}$$

where pf is the fluid pressure at the contacting HF tip. As a result of the NF activation due to the fracturing, the fluid penetration becomes very active in highly permeable NFs or with low viscosity fracturing fluids. This could potentially prevent the HF from propagation across the weak interfaces. The elasticity model of the fracture interaction enables the computation of the stress field in the vicinity of the activated NF. The analysis of the generated stress field gives the positions of sufficient tensile stress concentration where the new fractures can be nucleated. These positions in most situations correspond to the two opposite tips of the NF open zone (see Figure 2, right). In order to decide on the possibility of a secondary fracture (SF) re-initiation at these points, a criterion of fracture initiation which combines both stress criterion and energy release rate has been employed. The stress criterion requires that the maximum tensile hoop stress σθθ in the vicinity of the stress concentration point xjhaving direction θj with respect to the orientation of the NF must exceed the tensile strength of the rock T0 along the distance δT

$$\sigma_{\theta\theta}(x_j, r, \theta_j)_{r<\delta_T} \leq -T_0 \tag{3}$$

In addition, the energy criterion states that the elastic energy release rate due to the incremental initiation of a fracture of length δl must overcome the critical energy release rate for the given rock

$$\Im_{inc}(\delta l) > \Im_{1C}, \quad \delta l < \delta_T \tag{4}$$

The length of the fracture must not exceed the critical stress zone, δT. The mixed stress-energy criterion has been verified experimentally [26]. The model of HF-NF re-initiation has been validated against the results of various laboratory block tests [11, 12, 15]. The predictions of the analytical model for crossing and arresting behavior agree with the experimental results for various fracture intersection angles, stress contrasts and fluid injection conditions used in different experimental groups. Figure 3 shows the comparison between different analytical models [13, 14, 21], and the experimental results from [15]. The experiments clearly show that the new model agrees with the experiments as well as other analytical models as it captures the first order crossing-arresting behavior. We note that the discrimination between the different models would require additional data points in the transition zone, unfortunately not available here. Additionally, it should be noted that the injection rate and viscosity were not changed in this series of experiments, and so it was not possible to assess their effect on the fracture interaction outcome. In order to compensate for this lack of lab experiments, numerical experiments were conducted using MineHF2D code [4, 5, 8] to assess the sensitivity of the injection rate on fracture crossing. The results are demonstrated on Figure 4 and show that the OpenT model [20,21] agrees well with numerical computation results in the sense that it captures the crossing-arresting transition. It should be mentioned that the OpenT incorporates the influence of rock properties (local horizontal stresses, rock tensile strength, toughness, pore pressure, Young's modulus, Poisson ratio), natural fracture properties (friction coefficient, toughness, cohesion, permeability), intersection angle between hydraulic and natural fractures, fracturing fluid properties (viscosity, tip pressure), and injection rate to define crossing rules [21]. The eRP criterion [14,15] accounts for the local stress field, pore pressure, crossing angle, rock tensile strength and frictional properties of the natural fractures.

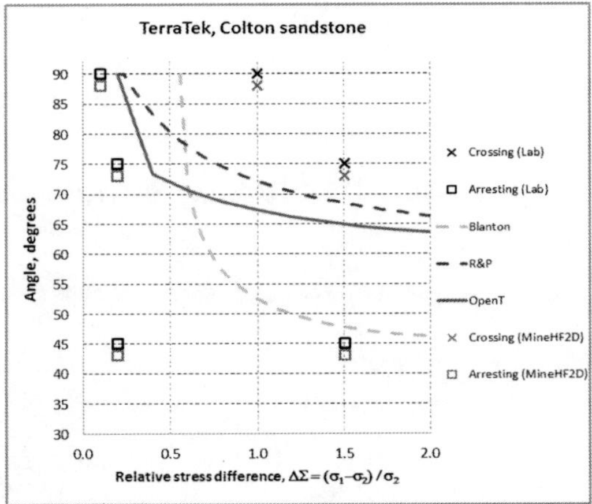

FIGURE 3. Comparison between analytical models given in [13,14,15], OpenT [21], and the experimental results [15]

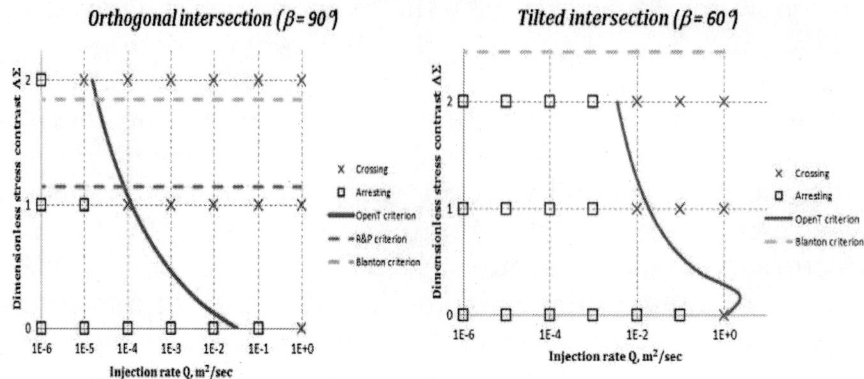

FIGURE 4. Comparison between numerical crossing-arresting HF-NF behavior using MineHF2D code [4,5,8]. The red crosses and squares respectively indicate crossing and arresting behavior from MineHF2D code, solid green curves correspond to analytical predictions using OpenT [21], dash yellow curve corresponds to Blanton criterion [13], and eRP criterion [14,15] is given by dash blue lines. The interaction is studied for various injection rate and relative stress difference for two different HF-NF contact angles, β=90° (left) and β=60° (right).

This new model has been implemented in UFM. Below we present comparison of UFM results with new and old crossing models and provide some analysis about the influence of fluid properties on the geometry of stimulated fracture network, and as a result on the production predictions.

Comparison of hydraulic fracturing simulations with OpenT versus eRP

INFLUENCE OF VISCOSITY

The comparison of results generated using two crossing criteria - eRP criterion [14,15] and new OpenT criterion [20, 21] - is presented in Figure 5 and Figure 6 for a simple example given in Table 1(values shown in italic are used only in OpenT crossing criterion). The cohesion and toughness of natural fracture are considered to be negligible.

TABLE 1.Input data Example 1

Injection rate	0.13 m^3/s
Stress anisotropy	0.9 MPa
Young's modulus	2.8 $\times 10^{10}$ Pa
Poisson's ratio	0.2
Fluid viscosity	0.001-0.01 Pa-s
Fluid Specific Gravity	1.0
Fracture toughness	1.3 MPa-$^{m0.}$5
Tensile strength	3.5 MPa
NF friction Coefficient	0.5
NF permeability	1 Darcy

For the case of lower fluid viscosity (Figure 5a and Figure 6a) both criteria show similar hydraulic fracture patterns with no crossing of the natural fractures. For higher viscosity fluid OpenT crossing criterion shows that hydraulic fractures cross the NF#1 and NF#3 (Figure 6b), while with eRP the hydraulic fracture network (HFN) pattern does not change. The intersection angle between HF and NF#1 was 62.5 deg, between HF and NF#2 was 15 deg, and the interaction angle between HF and NF#3 was 75 deg.

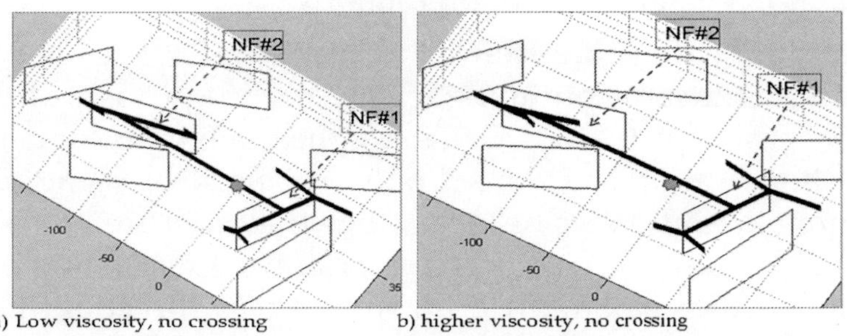

a) Low viscosity, no crossing b) higher viscosity, no crossing

FIGURE 5.Hydraulic fracture networks generated for Example 1 with eRP crossing criterion [14] with fluid viscosity K'=0.001Pa-s (left), and K'=0.01Pa-s (right)

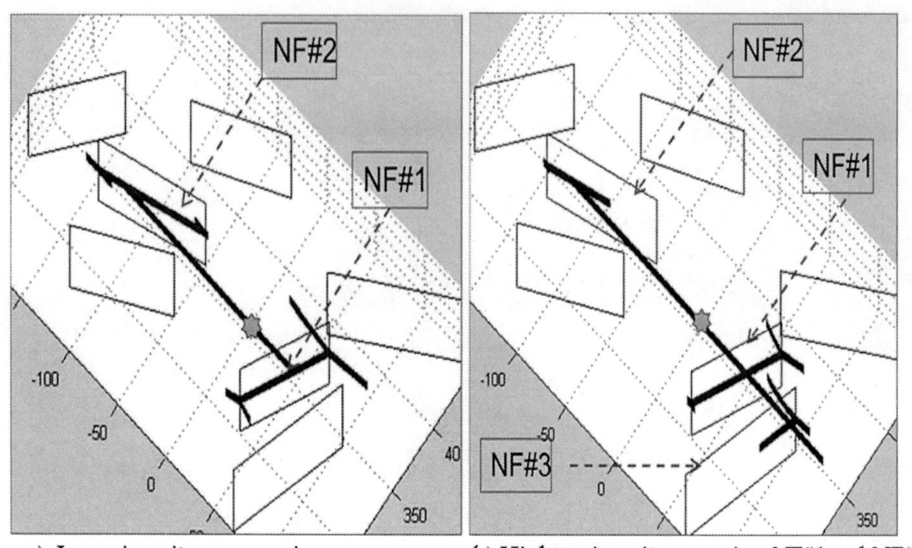

FIGURE 6.Hydraulic fracture networks generated for Example 1 with OpenT crossing criterion [21] with fluid viscosity K′=0.001pa-s (left) and K′=0.01Pa-s (right)

So, while eRP criterion gives for this case the same prediction (no crossing) for both low and high viscosity fluids, OpenT criterion predicts crossing the NF with higher crossing angle for the more viscous fluid.

Differences in the predicted hydraulic fracture network result in different prop pant placement (Figure 7), and will result in differences in production evaluation and prediction.

Example 2 with more dramatic output differences is presented in Figure 8 for the same pumping schedule, zone properties, fluid and natural fractures properties. In Table 2 the main input data is shown (values shown in italic are used only in OpenT crossing criterion), the toughness and cohesion of natural fractures are considered to be negligible. Natural fractures are oriented mostly perpendicular (~ 90deg) to the maximum horizontal stress direction, i.e. to the preferred direction of hydraulic fracture propagation.

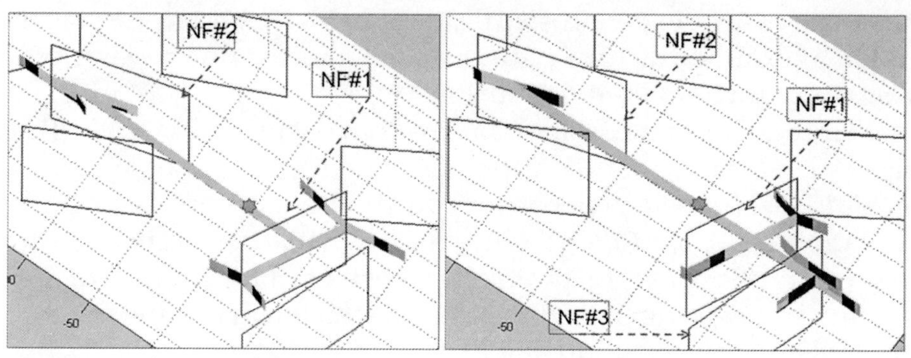

FIGURE 7. Proppant placement prediction for Example 1 from eRP (left) and OpenT (right) criteria with fluid K'=0.01Pa-s after 100 min of shut-in. Slurry is shown in light blue, bank is in dark blue, and clean fluid is in orange.

TABLE 2. Input data for Example 2

Injection rate	0.13 m^3/s
Stress anisotropy	2 MPa
Young's modulus	3.5×10^{10} Pa
Poisson's ratio	0.25
Fluid viscosity	0.0004-0.04 Pa-s
Fluid Specific Gravity	1.0
Min horizontal stress	42.7 MPa
Max horizontal stress	44.6 MPa
Fracture toughness	1 MPa-^{m0}.5
Tensile strength	3.4 MPa
NF friction Coefficient	0.4
NF permeability	1 Darcy

For the case of low viscosity fluid (Figure 8 left) both criteria show similar hydraulic fracture patterns with mostly no crossing of the natural fractures. When fracturing fluid viscosity was increased, considerable differences in patterns have been observed (Figure 8 right). The results for eRP approach stay mostly the same, showing that fluid eventually penetrated into the NF and opens it. But the simulation based on OpenT criteria shows that for higher viscosity fluid hydraulic fracture intersects most of the natural fractures, resulting in a bi-wing like HFN pattern, which will produce a narrow micro seismic events cloud.

The example presented on Figure 8b is consistent with the general observations that hydraulic fracturing treatments with higher fluid viscosity HFN tend to cross natural fractures and generate a narrower fracture network, while for low viscosity fluids it is easier to penetrate into the natural fracture and open it [9] and generate a wider fracture network.

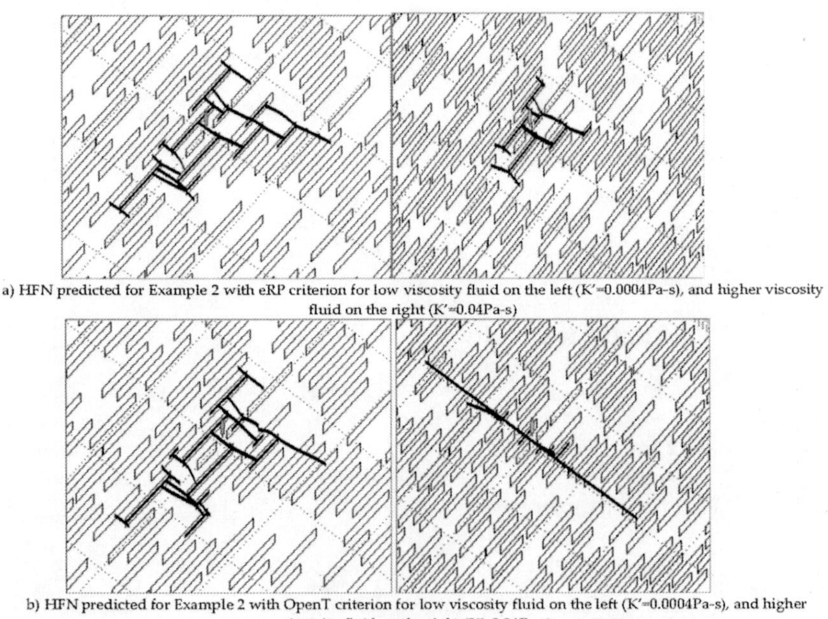

a) HFN predicted for Example 2 with eRP criterion for low viscosity fluid on the left (K'=0.0004Pa-s), and higher viscosity fluid on the right (K'=0.04Pa-s)

b) HFN predicted for Example 2 with OpenT criterion for low viscosity fluid on the left (K'=0.0004Pa-s), and higher viscosity fluid on the right (K'=0.04Pa-s)

FIGURE 8.Hydraulic fracture networks generated for Example 2 with eRP crossing criterion (a) and OpenT crossing criterion (b) for low and high fluid viscosity cases. The pre-existing DFN is also shown

It should be mentioned that rock properties, crossing angle between natural fracture and hydraulic fracture, and natural fractures properties all work together with fluid properties to define the crossing pattern and the resulting fracture footprint. This paper intends to emphasize the importance of fluid properties to be included into the general consideration for HF/NF interaction prediction.

INFLUENCE OF PUMPING RATES

As it was mentioned before, injection rate works together with fluid viscosity when HF interacts with NF [9, 18, 19]. In the OpenT crossing model the injection rate is also taken into account to predict (evaluate) HF/NF crossing or opening (Equation 1).

Notice that while using eRP criterion, the change in pumping rate can change fracture footprint due to change in fluid pressure, width, and therefore local stresses and crossing angle, while OpenT model introduces additional change due to the rate effect on the crossing behavior.

The cases presented in Examples 1-2 above, demonstrated the impact of fluid viscosity. Now these examples will be considered again to demonstrate the impact of pumping rate, which is also accounted for in the new crossing model. The base case of pumping rate Q=0.132 m3/s is considered and compared with additional cases when rate is changed (Table 3). The total pumping time in schedule was changed accordingly for different pumping rates to maintain the same total fluid volume.

TABLE 3. Input data to test impact of pumping rate

Fluid viscosity	0.0004-0.04 Pa-s
Injection rate : Q	0.132 m^3/s
Injection rate: Q/2	0.066 m^3/s
Injection rate: 5Q	0.66 m^3/s

First, on Figures 9a-10a the results of using the eRP crossing model with different rates as given in Table 3, and two types of fluid viscosity are presented for Example 1 and compared with the same simulations using OpenT crossing model (Figures 9b-10b). Due to relatively high leak off coefficient used in the presented case the fracture network for higher rate is larger due to greater fluid efficiency.

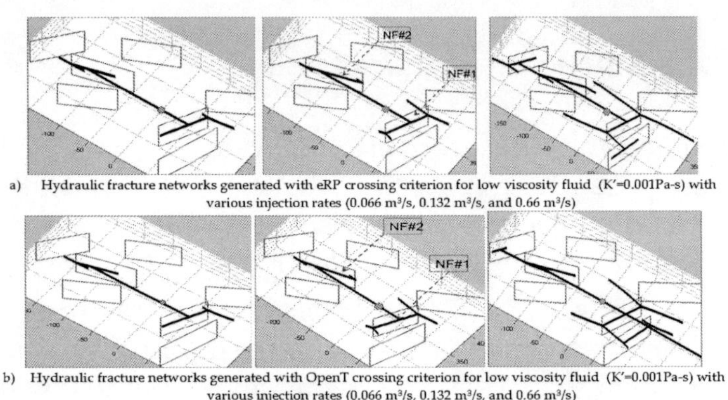

a) Hydraulic fracture networks generated with eRP crossing criterion for low viscosity fluid (K'=0.001Pa-s) with various injection rates (0.066 m^3/s, 0.132 m^3/s, and 0.66 m^3/s)

b) Hydraulic fracture networks generated with OpenT crossing criterion for low viscosity fluid (K'=0.001Pa-s) with various injection rates (0.066 m^3/s, 0.132 m^3/s, and 0.66 m^3/s)

FIGURE 9. Influence of Pumping Rate: Hydraulic fracture networks generated for Example 1 with low viscosity fluid and with eRP (a) and OpenT(b) crossing models at injection rates from Table 3

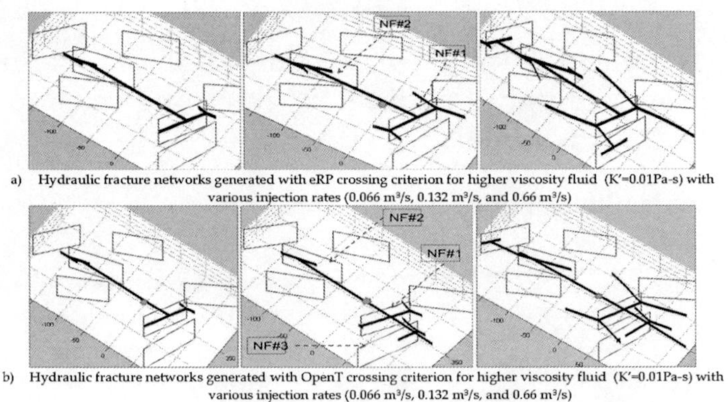

a) Hydraulic fracture networks generated with eRP crossing criterion for higher viscosity fluid (K'=0.01Pa-s) with various injection rates (0.066 m^3/s, 0.132 m^3/s, and 0.66 m^3/s)

b) Hydraulic fracture networks generated with OpenT crossing criterion for higher viscosity fluid (K'=0.01Pa-s) with various injection rates (0.066 m^3/s, 0.132 m^3/s, and 0.66 m^3/s)

FIGURE 10. Influence of Pumping Rate: Hydraulic fracture networks generated with higher viscosity fluid for Example 1 with eRP (a) and OpenT(b) crossing models at injection rates from Table 3

The first observation is that for low injection rate and for both high and low fluid viscosities fracture network is similar with both crossing models for this simple Example 1, and no crossing is observed. When injection rate is increased, HFN becomes more complex: OpenT shows crossing at the first natural fracture at the angle of 62.5 deg with both low and high viscosity fluids. The eRP criterion does not show crossing, and network complexity is due to the smaller time required to open NF and higher injection rates, so HFN can propagate faster. Again, the resulting HFN with eRP model does not depend on the fluid viscosity (Figure 9a and Figure10a). Results for the test case of Example 2 are given in Figures 11-12.

As we can see from Figures 11a and 12a, eRP criterion exhibits similar HFN footprints for both low and high viscosity fluids and for different injection rates. The reason for the relative insensitivity to the injection rate as compared to Example 1 is due to the higher stress anisotropy for Example 2. With OpenT crossing criterion, for lower fluid viscosity the chance of crossing perpendicular NF increases with increasing injection rate. At the same time for a higher viscosity fluid, while it can cross the natural fracture more easily, it is more difficult to open the crossed natural fracture. The observed behavior with new crossing model is consistent with experimental observations [9].

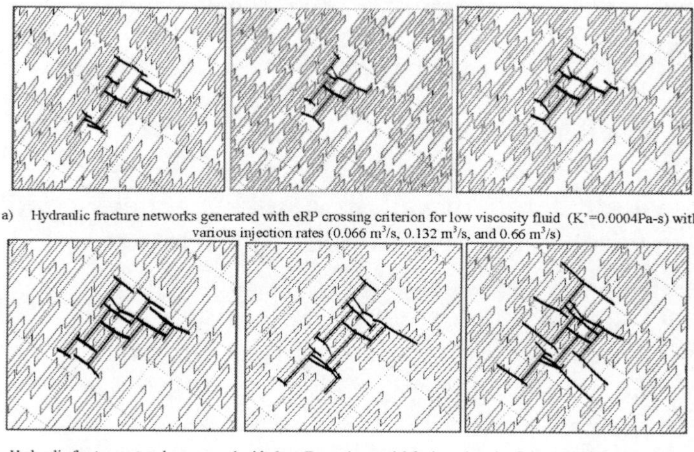

a) Hydraulic fracture networks generated with eRP crossing criterion for low viscosity fluid (K'=0.0004Pa-s) with various injection rates (0.066 m^3/s, 0.132 m^3/s, and 0.66 m^3/s)

b) Hydraulic fracture networks generated with OpenT crossing model for low viscosity fluid (K'=0.0004Pa-s) with various injection rates (0.066 m^3/s, 0.132 m^3/s, and 0.66 m^3/s)

FIGURE 11. Influence of Pumping Rate: Hydraulic fracture networks generated with low viscosity fluid for Example 2 with eRP (a) and OpenT(b) crossing models at injection rates from Table 3

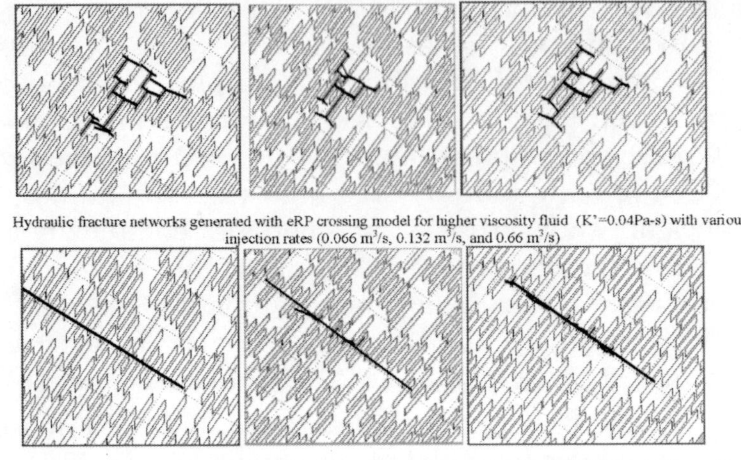

FIGURE 12. Influence of Pumping Rate: Hydraulic fracture networks generated for Example 2 with old and new crossing models for high viscosity fluid at injection rates from Table 3

So we can conclude from the observations in these cases that the fluid viscosity together with pumping rate could play a major role on the crossing. At the same time the influence of pumping rate is not as strong as viscosity, and mostly affects the opening of the intersected natural fractures.

BARNETT EXAMPLE

To further validate the model in a realistic field condition, we examine a synthetic case that mimics the field example in Barnett Shale presented by Warpinski et al. [16] as shown in Figure 1. Though the details of the well and formation data and pumping schedule are not exactly replicated, the synthetic case is created using the data that is available in [16], so the well and formation configurations are very close to the real case.

Some of the critical information for fracture simulation, including Young's modulus and description of the natural fractures, is prescribed based on the work by Gale et al. [27]. According to [27], the Young's modulus for Barnett Shale is 33 GPa (4.8 x 106 psi). The natural fractures contain a dominant set trending West-Northwest direction (approximately North 70° West). There is also another set trending North-South direction. The hydraulic fractures in Barnett trend in the Northeast-Southwest direction. The natural fractures are mostly sealed and filled with calcite. Only largest fractures may be open and largest fracture clusters are expected to space couple hundred feet apart. To construct the natural fractures for UFM simulation, we assume that only the largest natural fractures contribute to the complex fracture network development. The exact values of fracture spacing and fracture length are difficult to determine. We make the assumption that the average fracture spacing is 100 ft and average fracture length is 200 ft. Only the dominant set of fractures is assumed. Figure 13 shows the top view of the well configuration, perforation clusters and the 2D traces of the generated natural fractures. The well geometry closely mimics the field case as shown in Figure 1.

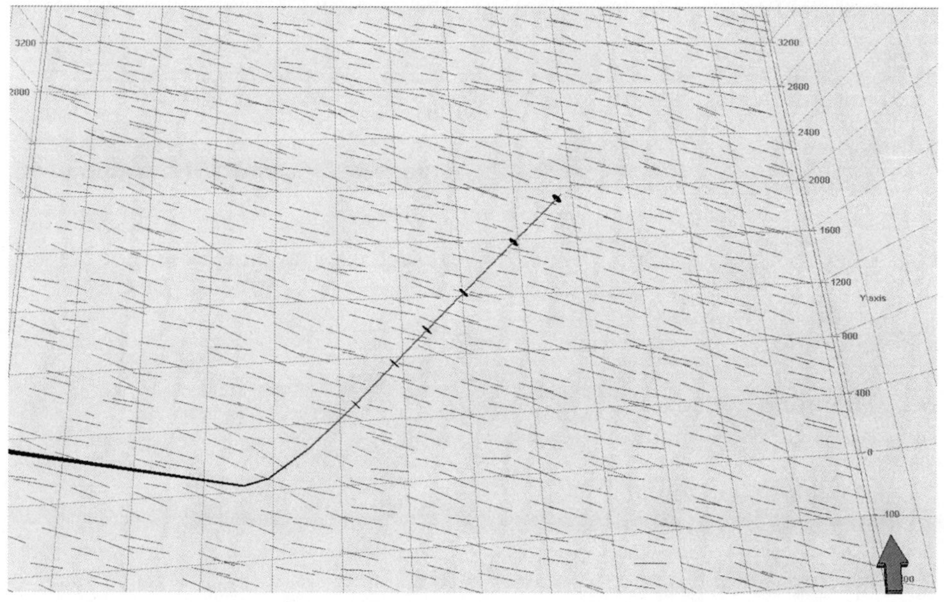

FIGURE 13. Top view of the wellbore, perforations and the natural fractures used for the Barnett simulations

For the complex fracture simulation, detailed vertical stress profile is not available from [16]. Instead a fixed height model is used based on the micro seismic measurements presented in [16]. It is assumed that the fracture height is 310 ft covering Lower Barnett for the case of cross-linked gel treatment, and 360 ft for the slick water treatment. For the simulation of slick water refrac, any potential effect of previous cross-linked treatment and the small slick water treatment prior to the main treatment is not considered. Furthermore, a difference between maximum and minimum horizontal stress is assumed to be 200 psi. Table 4 shows the main parameters used for the fracture simulations.

TABLE 4. Input data for Barnett Shale case

Parameters	Xlink Gel treatment	Slick Water treatment
Young's modulus	4.8 x 10^6 psi	
Natural fracture direction	Average N70°W, standard deviation 5°	
Natural fracture length	Average 200 ft, standard deviation 40 ft	
Natural fracture spacing	Average 100 ft, standard deviation 20 ft	
Coefficient of friction	Average 0.6, standard deviation 0.1	
Hydraulic fracture direction	N40°E	
Minimum horizontal stress	5324 psi	
Maximum horizontal stress	5524 psi	
Fracture height	310 ft	360 ft
Fluid rheology	n′ = 0.42, k′ = 0.002 lb-s/f^t2	1 cp
Injection rate : Q	70 bpm	125 bpm
Pump time	174 min	386 min

Parameters	Xlink Gel treatment	Slick Water treatment
Proppant volume	715,000 lbs	600,000 lbs

Figure 14 shows the UFM simulated fracture geometry and width for both gel and slick water fracs at the end of the treatments.

Planar hydraulic fractures first initiate from the perforations. These fractures propagate as longitudinal fractures since the wellbore direction is closely aligned with the fracture orientation. For the cross-linked gel treatment, as these initial longitudinal fractures intersect the natural fractures that are approximately orthogonal to the fracture direction, the OpenT crossing model mostly predicts crossing through the natural fractures.

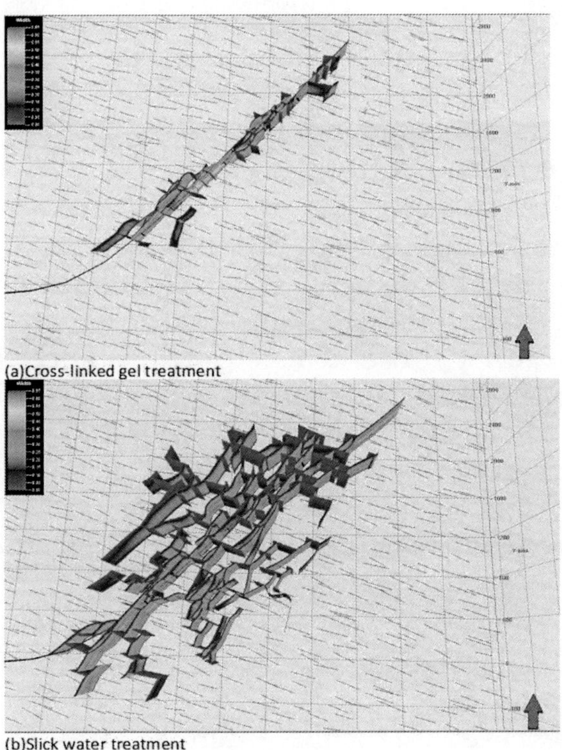

FIGURE 14. UFM simulation results for the Barnett case

Only when the fluid pressure is sufficiently high to exceed the normal stress acting on the natural fractures, do the natural fractures be opened up and accept fracturing fluid. The overall geometry predicted by UFM model shows a strong planar trend along the well with very narrow network width, consistent with the micro seismic observation shown in Figure 1a.

For the slick water treatment, the OpenT crossing model mostly predicts non-crossing condition when a hydraulic fracture intercepts a natural fracture. This results in a much wider fracture network width as the fractures branch out as shown in Figure 14b. The width of the network is approximately 1700 ft wide, approximately the same as indicated by the micro seismic data as shown in Figure 1b.

Example presented on Figure 14 showing the difference in HFN from two treatments with different types of fluid, matches micro seismic cloud trend observed in [16] and definitely shows ability of UFM simulator with new implemented crossing model correctly predict hydraulic fracture complexity in naturally fractured formation.

Possible impact on production forecast

Unconventional Fracture Model (UFM) presents a powerful tool to evaluate hydraulic fracture network propagation under the specified field pumping conditions and can be used to predict developed hydraulic fracture network and match it with observed micro seismic event cloud. The proper understanding of the fracture footprint as well as propped fracture surface estimation is an important input for the production evaluation. Because the OpenT crossing model predicts some changes in crossing patterns with different fluids used, it is important to understand the impact it could have on the production evaluation.

This section illustrates how the crossing criterion can influence the production. It presents simulation results of a fracturing-to-production simulation. The production part is done with the UPM model [28]. The base case for this Example 3 is from the paper [29] (Table 5).

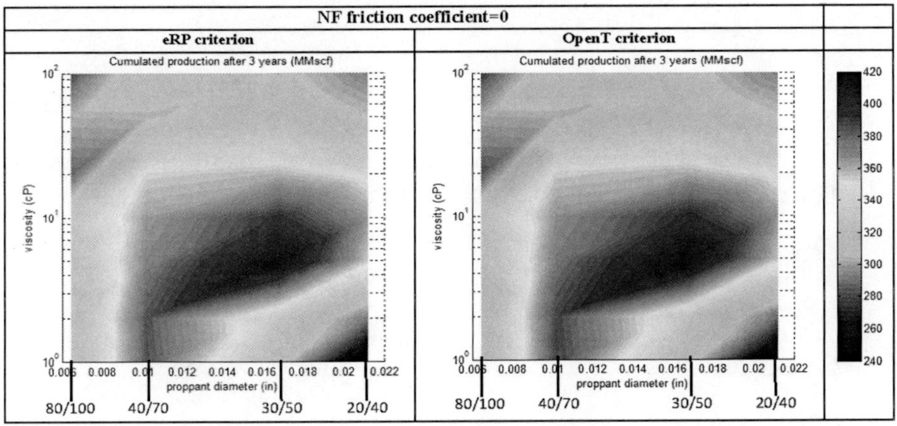

FIGURE 15. Cumulated production after 3 years for the two crossing criterion and a friction coefficient at NF = 0, as a function of the fracturing fluid viscosity and the prop pant size

Figure15 shows the cumulated production after 3 years as a function of the prop pant size and the fracturing fluid viscosity for the case when zero friction coefficient at NF is used and the two crossing models are applied. From Figure 15 we see that if there is no friction at the natural fractures, there is no difference between results from the two crossing criteria for this case. The reason is that if friction coefficient for the natural fracture is zero, both crossing criteria show that HF will not cross NF. The HFN footprint and fracture conductivity (identical when using both crossing criteria with zero friction coefficient) are shown at Figure 16 a,b for cases of slick water and more viscous fluid pumped.

TABLE 5. Input data for Example 3

Injection rate	0.21 m^3/s
Number of Perforated Intervals	4
Stress anisotropy	0.3 MPa
Young's modulus	1.3×10^{10} Pa
Poisson's ratio	0.23

Fluid viscosity	0.001-0.1 Pa-s
Min horizontal stress	28.47 MPa
Max horizontal stress	28.76 MPa
Fracture toughness	1.5 MPa-$^{m0.}$5
Tensile strength	3.4MPa
NF friction Coefficient	0 -0.75
NF permeability	1 Darcy

If HF cannot cross NF, it is easier for slick water to penetrate and open NFs than for more viscous fluid, so HFN is generally more extended (Figure 16a left). While for the case of more viscous fluid it takes more time to open NF, HFN pattern is smaller (Figure 16)

When friction coefficient at NF is increased from 0 to 0.75, the output changes depending on the type of fluid pumped. The cumulated production forecast looks different for two crossing criteria used (Figure17).

For slick water treatments (Figure 18), eRP criterion shows some crossing of NFs (Figure18 left), while OpenT claims that no crossing should occur for slick water treatments (Figure 18 right). This difference in crossing models, produces considerable differences in production prediction after 3 years for low viscosity fluid treatments (Figure 17). It is important to mention, that it is a common observation that low viscosity fluids usually do not cross NFs, mainly because it is easier for them to penetrate to NF and open it [9]. The eRP criterion cannot capture this effect, while OpenT model correctly predicts HF/NF interaction for slick water case with friction coefficient at NF of 0.75.

When more viscous fluid pumped, results also show some differences (Figure19). With both criteria some crossing is observed, but OpenT in this case predicts more crossing than eRP criterion. Mention, that differences in results from eRP criterion for slick water and 100cP fluid are due to some differences in interaction angles between HF and NFs due to change in fluid properties, fluid pressure and due to stress shadow effect.

Due to small stress field anisotropy, the differences in production prediction for more viscous fluid are not significant (Figure 17), but still visible.

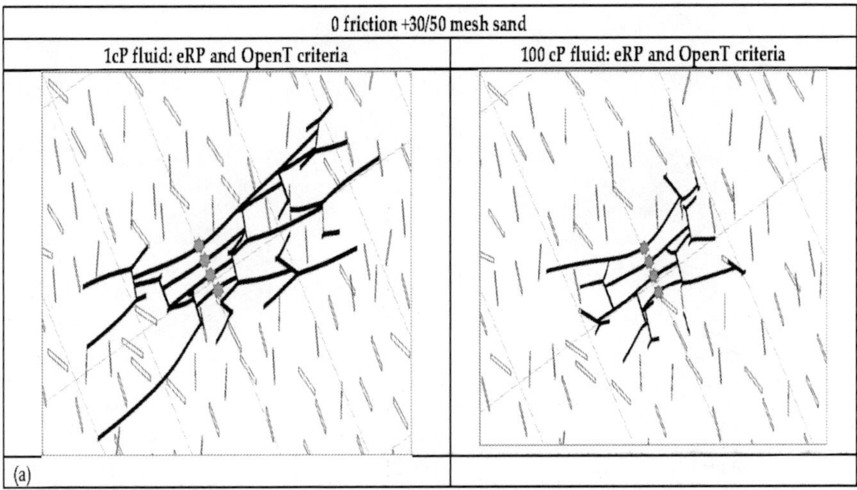

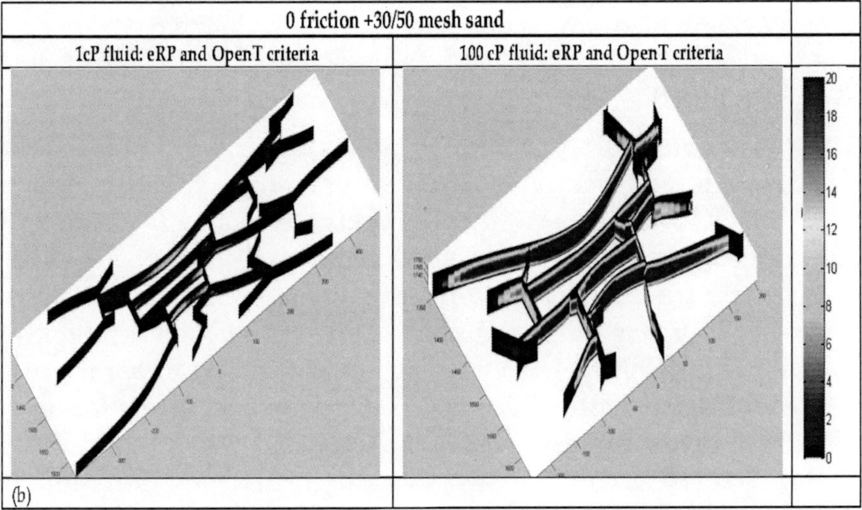

FIGURE 16.(a). HFN footprint for two types of fluid (with 30/50 mesh sand) for zero friction coefficient at NF. Both criteria show the same HFN footprint. (b). Fracture conductivity for two types of fluid (with 30/50 mesh sand) for zero friction coefficient at NF. Both criteria show the same HFN footprint

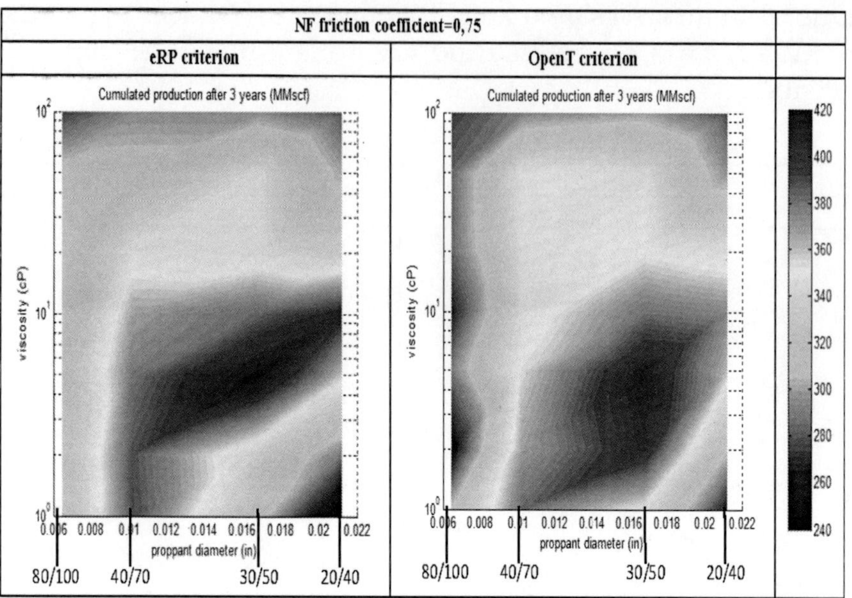

FIGURE 17. Cumulated production after 3 years for the two crossing criteria and a friction coefficient at NF of 0.75, as a function of the fracturing fluid viscosity and the prop pant size

The main conclusion related to presented production examples, is that the difference between the two crossing criteria seems to be maximum for low viscosity fluid (slick water) and large prop pant (30/50). This observation is expected because the lower the viscosity, the longer the fracture length and the stronger interaction with NF are. Also, eRP criterion shows some crossing of NFs for slick water case, while OpenT shows no crossing, and larger prop pants are more sensitive to fracture intersections. The fracture width is larger if the HF does not cross NF and slurry propagated inside the NF with a larger normal stress (in case of stress anisotropy) and smaller width, thus increasing the likelihood of bridging. Also, the less crossing occurs, the more time HF needs to spend stopped at NF before building enough pressure to overcome the stress anisotropy and resume propagating inside the NF for high viscosity fluids. In this case, more prop pant will settle close to the perforations, reducing the propped length and thus the production.

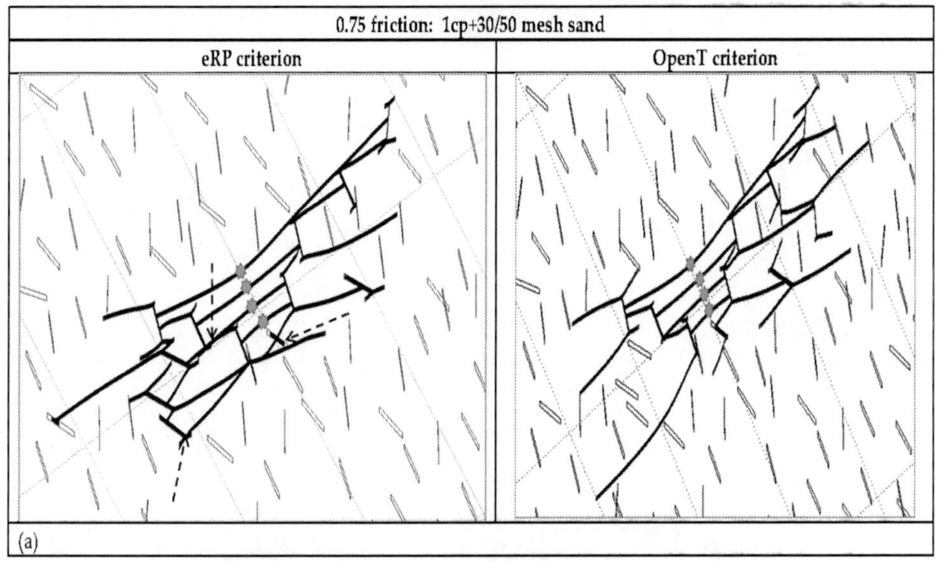

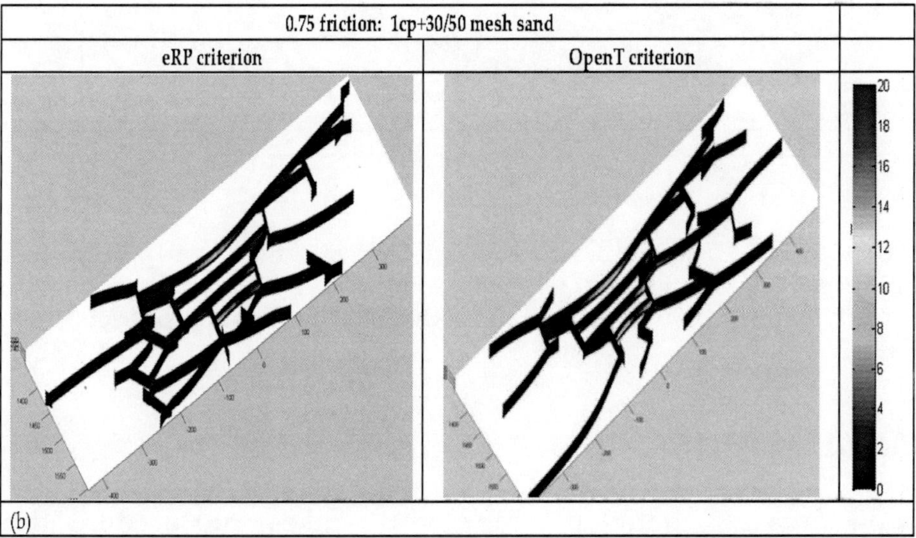

FIGURE 18.(a). HFN footprint for slick water (with 30/50 mesh sand) for friction coefficient at NF=0.75. Both criteria show similar HFN footprint. eRP shows some crossing (shown by dashed arrows), while OpenT shows no crossing for slick water case (b). Fracture conductivity for slick water (with 30/50 mesh sand) for friction coefficient at NF=0.75

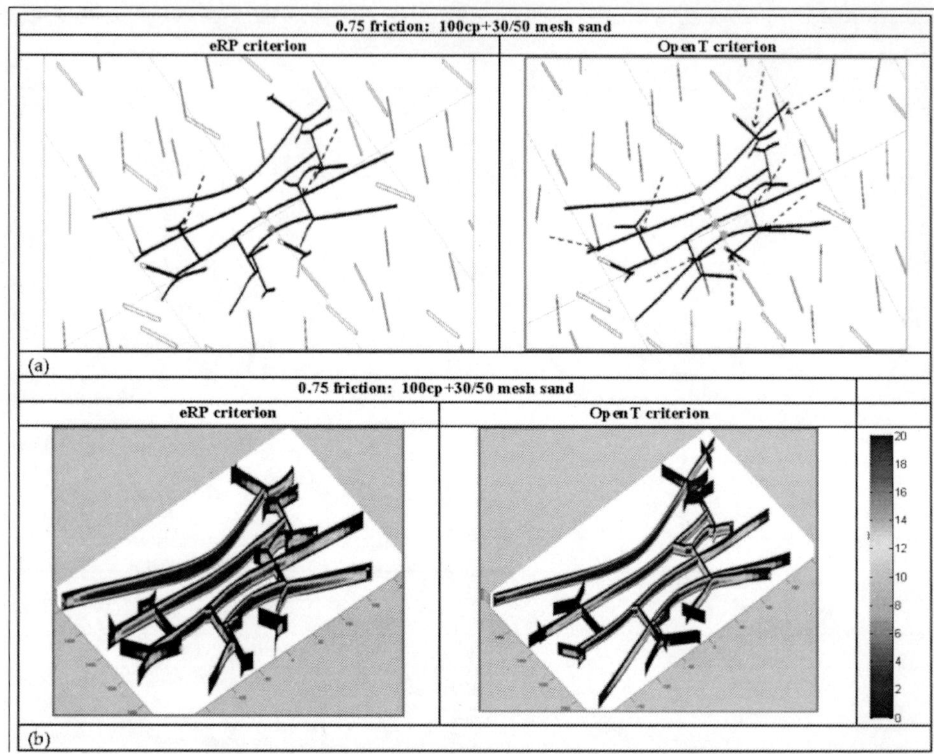

FIGURE 19.(a). HFN footprint for 100cP viscosity fluid (with 30/50 mesh sand) for friction coefficient at NF=0.75. Arrows point at crossing (b). Fracture conductivity for 100cP viscosity fluid (with 30/50 mesh sand) for friction coefficient ant NF=0.75

CONCLUSIONS

A new crossing model (OpenT) which takes into account fluid properties, properties of the rock mass and natural fractures, have been developed, validated [20, 21], and implemented in UFM. The similarities and differences in fracture footprint predicted based on the OpenT model and eRP criterion have been demonstrated and discussed. OpenT crossing model shows more realistic results for some cases (as for the field and laboratory observations) than existing purely rock property based models.

While eRP model properly accounts for interaction angle, stress anisotropy, rock tensile strength and NF friction coefficient, the OpenT model accounts also for NF and fluid properties. The crossing prediction from OpenT criterion, and therefore corresponding HFN footprint, could be different for low viscosity fluids and high viscosity fluids, while eRP model shows similar crossing patterns for low and high viscosity fluids. In the meantime both criteria show similar results for some cases.

It is important to mention that whether HF will cross (dilate, or open) NF depends on the combined impact of rock properties (local stress field, tensile strength, toughness, etc), NF properties (permeability, toughness, friction coefficient, cohesion, etc), HF/NF interaction angle, fluid properties, injection rate and other properties.

In general, the Unconventional Fracture Model (UFM) with new OpenT crossing model provides more reliable results to predict and evaluate hydraulic fracture network geometry and improve production forecast. The Barnett Shale example presented in Figure 14 shows the differences in HFN from two treatments with two different types of fluids. The predicted results closely match the micro seismic cloud observed in [16] and show the ability of UFM simulator with the new crossing model (OpenT) to correctly predict hydraulic fracture complexity in naturally fractured formation.

ACKNOWLEDGEMENTS

The authors would like to thank Schlumberger for permission to present and publish this paper

REFERENCES

1. Chuprakov DS, Akulich AV, Siebrits E, Thiercelin M. Hydraulic Fracture Propagation in a Naturally Fractured reservoir. SPE 128715, Presented at the SPE Oil and Gas India Conference and Exhibition held in Mumbai, India, 20-22, January, 2010.
2. Zhao J, Chen M, Jin Y, Zhang G. Analysis of fracture propagation behavior and fracture geometry using tri-axial fracturing system in naturally fractured reservoirs. Int. J. Rock Mech. & Min. Sci 2008; 45: 1143-1152.

3. Thiercelin M, Makkhyu E. Stress field in the vicinity of a natural fault activated by the propagation of an induced hydraulic fracture. In: Proceedings of the 1st Canada-US Rock Mechanics Symposium 2007; 2:1617-1624.
4. Zhang X, Jeffrey RG. The role of friction and secondary flaws on deflection and re-initiation of hydraulic fractures at orthogonal pre-existing fractures. Geophys J Int. 2006;166(3): 1454-1465.
5. Zhang X, Jeffrey RG. Reinitiation or termination of fluid-driven fractures at frictional bedding interfaces. J Geophys Res-Sol Ea. 2008; 113(B8), B08416.
6. Zhang X, Jeffrey RG, Thiercelin M. Effects of Frictional Geological Discontinuities on hydraulic fracture propagation. SPE 106111, Presented at the SPE Hydraulic Fracturing Technology Conference, College Station, Texas, January29-31, 2007.
7. Zhang X, Jeffrey RG., Thiercelin M. Deflection and propagation of fluid-driven fractures as frictional bedding interfaces: a numerical investigation. Journal of Structural Geology 2007; 29: 390-410.
8. Zhang X, Jeffrey RG, Thiercelin M. Mechanics of fluid-driven fracture growth in naturally fractured reservoirs with simple network geometries. Journal of Geophysical Research 2009;114, B12406.
9. Beugelsdijk LJL, de Pater CJ, Sato K. Experimental hydraulic fracture propagation in a multi-fractured medium. SPE 59419, Presented at the SPE Asia Pacific Conference in Integrated Modeling for Asset Management, Yokohama, Japan, April 25-26, 2000.
10. Renshaw CE, Pollard DD. An Experimentally Verified Criterion for Propagation across Unbounded Frictional Interfaces in Brittle, Linear Elastic-Materials. International Journal of Rock Mechanics and Mining Sciences & Geomechanics Abstracts. 1995 Apr, 32(3): 237-49.
11. Warpinski NR., Teufel LW. Influence of Geologic Discontinuities on Hydraulic Fracture Propagation (includes associated papers 17011 and 17074). SPE Journal of Petroleum Technology 1987;39(2): 209-220
12. Blanton TL. An Experimental Study of Interaction Between Hydraulically Induced and Pre-existing Fractures. SPE 10847, Presented at the SPE/DOE Unconventional Gas Recovery Symposium, Pittsburgh, PA, May 16-18, 1982.
13. Blanton TL. Propagation of Hydraulically and Dynamically Induced Fractures in Naturally Fractured Reservoirs. SPE Unconventional Gas Technology Symposium; 01/01/1986; Louisville, Kentucky,1986.
14. Gu H, Weng X. Criterion For Fractures Crossing Frictional Interfaces At Non-orthogonal Angles. 44th US Rock Mechanics Symposium and

5th US-Canada Rock Mechanics Symposium; 01/01/2010; Salt Lake City, Utah: American Rock Mechanics Association; 2010.

15. Gu H, Weng X, Lund JB, Mack M, Ganguly U, Suarez-Rivera R. Hydraulic fracture crossing natural fracture at non-orthogonal angles, a criterion, its validation and applications. Paper SPE 139984 presented at the SPE Hydraulic Fracturing Conference and Exhibition, Woodlands, Texas, 24-26 January, 2011.
16. Warpinski NR, Kramm RC, Heinze JR, Waltman CK. Comparison of Single- and Dual-Array Microseismic mapping Techniques in the Barnett Shale. SPE 95568. Presented at the 2005 SPE Annual Technical Conference and Exhibition, Dallas, Texas, October 9012, 2005.
17. Cipolla C, Weng X, Mack M, Ganguly U, Gu H, Kresse O, Cohen C. Integrating Microseismic Mapping and Complex Fracture Modeling to Characterize Fracture Complexity. SPE 140185 Presented at the SPE Hydraulic Fracturing Technology Conference and Exhibition in The Woodlands, Texas, USA, January 24-26 2011.
18. Gil I, Nagel N, Sanchez-Nagel M. The Effect of Operational Parameters on Hydraulic Fracture Propagation in Naturally Fractured Reservoirs - Getting Control of the Fracture Optimization Process. Presented at 45th US Rock Mechanics/ Geomechanics Symposium, San Francisco, CA, 26-29 June, 2011.
19. Nagel N, Gil I, Sanchez-Nagel M, Damjanac B. Simulating Hydraulic Fracturing in Real Fractured Rock - Overcoming the Limits of Pseudo 3D Models. SPE 140480, presented at the SPE HFTC in Woodlands, Texas, USA, 24-26 January, 2011.
20. Chuprakov D, Melchaeva O, Prioul R. Hydraulic Fracture Propagation Across a Weak Discontinuity Controlled by Fluid Injection. InTech; 2013.
21. Chuprakov D, Melchaeva O, Prioul R. Injection-sensitive mechanics of hydraulic fracture interaction with discontinuities. Submitted for ARMA Symposium 2013.
22. Kresse O, Cohen C, Weng X, Wu R, Gu H. Numerical modeling of hydraulic fracturing in naturally fractured formations. 45th US Rock Mechanics/ Geomechanics Symposium, San Francisco, CA, 26-29 June, 2011.
23. Weng X., Kresse O., Cohen C., Wu R., Gu H. Modeling of Hydraulic Fracture Network Propagation in a Naturally Fractured Formation. Paper SPE 140253 presented at the SPE Hydraulic Fracturing Conference and Exhibition, Woodlands, Texas, USA, 24–26 January, 2011.
24. Kresse O, Weng X, Wu R, Gu H. Numerical modeling of Hydraulic

fractures interaction in complex Naturally fractured formations. ARMA-292, Presented at 46th US Rock Mechanics /Geomechanics Symposium, Chicago, Il, USA, 24-27 June 2012.

25. Valko P, Economides MJ. Hydraulic Fracture Mechanics: John Wiley & Sons; 1995.
26. Leguillon D. Strength or toughness? A criterion for crack onset at a notch. Eur J Mech a-Solid. 2002 Jan-Feb;21(1):61-72.
27. Gale JFW, Reed RM, Holder J. Natural fractures in the Barnett Shale and their importance for hydraulic fracture treatment. AAPG Bulletin, v91, No. 4: 603-622, April 2007.
28. Cohen CE, Xu W, Weng X, Tardy P. Production Forecast After Hydraulic Fracturing in Naturally Fractured Reservoir: Coupling a Complex Fracturing Simulator and a Semi-Analytical Production Model. SPE paper 152541 presented at the SPE Hydraulic Fracturing Technology Conference and Exhibition held in The Woodlands, Texas, USA, 6-8 February 2012.
29. Cohen CE, Abad C, Weng X, England K, Phatak A, Kresse O, Nevvonen O, Lafitte V, Abivin P. Analysis on the Impact of Fracturing Treatment Design and Reservoir Properties on Production from Shale Gas Reservoirs. IPTC 16400, Presented at the International Petroleum Technology Conference, Beijing, China, 26-28 March, 2013.

Chapter 11

HYDRAULIC FRACTURING IN FORMATIONS WITH PERMEABLE NATURAL FRACTURES

Olga Kresse[1] and Xiaowei Weng[1]

[1] Schlumberger, Sugar Land, USA

The recently developed Unconventional Fracture Model (UFM*) simulates complex hydraulic fracture network propagation in a formation with pre-existing closed natural fractures, and explicitly models hydraulic injection into a fracture network with multiple propagating branches [1]. The model predicts whether a hydraulic fracture front crosses or is arrested by a natural fracture it encounters, which defines the complexity of the generated complex hydraulic fracture network.

While taking into account the leak off of the fracturing fluid into the formation, the leak off into the natural fractures should also be considered, especially in low-matrix permeability conditions. The transmissibility of natural fractures can become significant, and the fracturing fluid can penetrate into natural fractures. Different regions can coexist along the invaded natural fracture: hydraulically opened region filled with fracturing fluid, region of still closed natural fracture invaded by fracturing fluid due to natural fracture permeability, and the region of natural fracture filled with original reservoir fluid.

Explicit modeling of hydraulic fractures interacting with permeable natural fractures becomes extremely complicated with the necessity to account for conservation of fluid mass, pressure drop along natural fractures, leak-off into the formation from natural fracture walls, pressure sensitive natural fracture permeability, properties of natural fractures, fluid rheology, while tracking the interface of each region along invaded natural fracture. A main challenge is integrating this hydraulic fracture/natural fracture interaction modeling into the overall hydraulic fracture network propagating scheme without losing model effectiveness and CPU performance.

The updated UFM model with enhancement to account for leak off into the natural fractures will be presented.

INTRODUCTION

It is believed that complexity of the resulting fracture network during hydraulic fracturing treatments in formations with pre-existing natural fractures is caused mostly by the interaction between hydraulic and natural fractures. Natural fractures can be important for hydrocarbon production in the majority of low-permeability reservoirs, particularly where the permeability of the rock matrix is negligible. Understanding and proper modeling of the mechanism of hydraulic-natural fractures interaction is a key to explain fracture complexity and the micro seismic events observed during HF treatments, and therefore to properly predict production.

When hydraulic fracture (HF) intercepts natural fracture (NF) it can cross the NF, open (dilate) the NF, or be arrested at NF. If hydraulic fracture crosses natural fracture, it remains planar, with a possibility to open the intersected NF if the fluid pressure at the intersection exceeds the effective stress acting on the NF. If the HF does not cross the NF, it can dilate and eventually propagate into the NF, which leads to more complex fracture network.

The interaction between HF and NF depends on in-situ rock stresses, mechanical properties of the rock, properties of natural fractures, and hydraulic fracture treatment parameters including fracturing fluid properties and injection rate. During the last decades, extensive theoretical, numerical, and experimental work has been done to investigate, explain, and use the rules controlling HF/NF interaction [3-17]. A new crossing model [2] recently implemented in UFM is able to predict the crossing behavior of HF at the NF accounting for the effects of fluid properties and NF permeability [18].

One of the important effects of natural fractures is enhanced leak off, which can lead to a premature screen out during proppant injection. In a formation with low-matrix-permeability, the transmissibility of natural fissures can be significantly higher than that of the reservoir matrix. The fracturing fluid can readily penetrate into natural fissures during the fracturing process and maintain a pressure nearly equal to the pressure in the primary fracture [19].

The concept that natural fractures (fissures) could alter leakoff has been a subject of numerous studies [20-24] with considering the fissure opening conditions or pressure-sensitive leakoff conditions. It was often reported that permeability of natural fractures is pressure dependent [23, 25,26].

The ways that elevated pressure could affect natural fractures have been described in [27]. Fissures with rough surfaces and minimal mineralization are most likely highly sensitive to the net stress pushing on them. Under virgin reservoir conditions (when the pressure p within the fissure equals the initial reservoir pressure p_{ini}), the effective stress is fairly high and the open channels formed by mismatched fracture faces are most likely deformed and nearly closed. As the pressure in the fissure increases because of leakoff of the high-pressure fracturing fluid ($p > p_{ini}$), the net closure stress is reduced and the fissure porosity opens. In this regime, the leakoff coefficient is highly pressure dependent. As the pressure exceeds the closure stress on the fissure ($p > p_{fo}$), the entire fissure opens, yielding an accelerated leakoff condition. The estimation of the critical pressure in PKN-type HF (exceeding closing pressure p > p fo) to open a vertical fissure has been given through the function of the principal horizontal stresses and Poisson ratio [20].

A more detailed description of the effects from natural fissures in reservoirs where natural fissures are the primary source of permeability is provided in [23]. The enhanced rate of fluid loss throughout the treatment is predicted, with leak off accelerating as the fracturing pressure increases. The increase in fluid pressure in the fissures reduces the effective normal stress acting to close the fissures and hence increases their permeability. For hydraulic fracturing purposes, the effect of the magnified permeability is reflected as an increase in fluid-leak off coefficient. The fluid-leak off in the presence of natural fissures could be as high as 2 to 3 times that for normally occurring pressure - dependent leak off behavior, even under the net pressure conditions.

For slightly elevated pressures NF porosity begins to open as the pore pressure increases because the elevated pressure relieves some of the net stress on the asperity contacts. Several models of this process have been developed. For example, [26] predicts the change in NF permeability resulting from changes in stress and pressure. This model have been validated and used in numerous studies [23].

Among existing HF models accounting for the permeability of intercepted natural fractures mention [10] which couples fluid flow, elastic deformation, and frictional sliding to obtain a solution which depends on the competition between fractures for the permeability enhancements. The effect of initially closed but conductive fracture is specifically addressed. The possible scenarios for evolution of fracture opening and fluid transport in closed NFs implemented in [10] are shown in Figure 1. The initial aperture w 0 along a closed pre-existing NF corresponds to its residual conductivity. It is equal to the effective aperture for the parallel plate model. The initial conductivity of a closed natural fracture arises from the fact that its surfaces are rough and mismatched at fine scale, i.e. the aperturew 0 is related to the fracture porosity. With increasing the fluid pressure, the hydraulic aperture will slightly change due to micro structural change in the natural fracture, although the fracture still remain closed and carries some contact stresses. In the end, fracture will be opened mechanically as the fluid pressure exceeds the normal stress acting on the fracture. In this case, the effective hydraulic conductivity is equal to the sum of both hydraulic aperture and mechanical opening since the fracture opening augments the initial hydraulic aperture, as shown in Figure 1. Zhang's model also considers the possibility of frictional sliding through the Coulomb frictional law, and accounts for three types of contact behavior at fracture surface: fracture is opened, fracture is closed but surface is in sticking mode, and fracture is closed but in sliding mode.

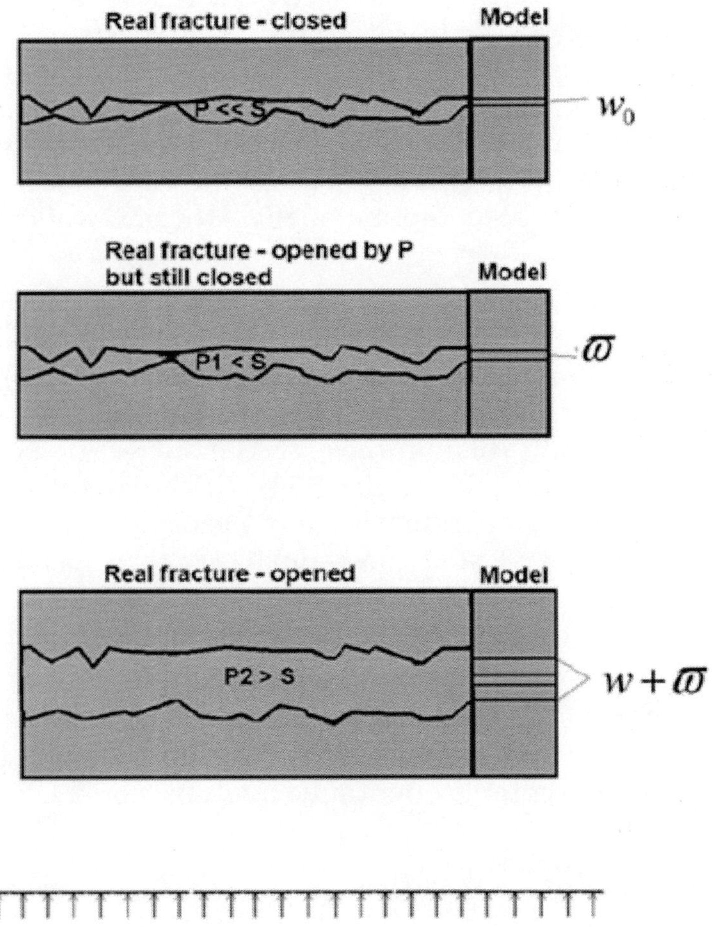

FIGURE 1. Evolution of natural fracture opening [10]

The HF models [28-32] do not account for permeability of natural fractures explicitly. The 2D model in [33] uses approach from [13] to simulate interaction between induced propagating fracture and natural fracture. A modified leak-off model for an intersecting fracture based on poro-elasticity was introduced to account for the increased leakoff at the intersections. A poro-elastic solution for the stresses in the HF/NF interaction zone has been used as a basis for hydraulic/natural fracture interaction criteria. A fully coupled finite element based approach was used to simulate HF propagation in a poroelastic formation with existing natural fractures.

The approach given in [10] is based on boundary element method and rigorously models HF interaction with permeable NF. It is computationally expensive, and is applicable for analysis of limited (small) number of HF/NF interactions. For a more general complex fracture network model like UFM which deals with a large number (order of thousands) of natural fractures, the CPU time is important and model should be computationally efficient while still being physically correct.

The important aspect of HF/NF interaction is shear slippage of NF faces. The possibilities of shear slippage in natural fractures due to change of stress field (in isolated natural fractures) or during HF/NF interactions, and the influence of shear slippage on fracture aperture change and dilation have been a subject of experimental and numerical studies [10, 15, 17, 30, 34-37]. The conditions for shear slippage and the corresponding shear displacement (apertures) have been investigated [36], and the estimation of permeability of NF with changing effective normal stress is done by [11,38]. The shear slippage effect during HF/NF interaction is also included in current approach.

This paper describes how leakoff into the natural fractures during HF/NF interaction (crossing or arresting before NF opens) is integrated into the complex hydraulic fracture model UFM.

UFM MODEL SPECIFICS

A complex fracture network model, referred to as Unconventional Fracture Model (UFM), had recently been developed [1, 39, 40]. The model simulates the fracture propagation, rock deformation, and fluid flow in the complex fracture network created during a treatment. The model solves the fully coupled problem of fluid flow in the fracture network and the elastic deformation of the fractures, which has similar assumptions and governing equations as conventional pseudo-3D fracture models. Transport equations are solved for each component of the fluids and proppants pumped. A key difference between UFM and the conventional planar fracture model is being able to simulate the interaction of hydraulic fractures with pre-existing natural fractures, i.e., determine whether a hydraulic fracture propagates through or is arrested by a natural fracture when they intersect and subsequently propagates along the natural fracture.

To properly simulate the propagation of multiple or complex fractures, the fracture model takes into account the interaction among adjacent hydraulic fracture branches, often referred to as "stress shadow" effect. It is well known that when a single planar hydraulic fracture is opened under a finite fluid net pressure, it exerts a stress field on the surrounding rock that is proportional to the net pressure. The details of stress shadow effect implemented in UFM are given in [40].

The branching of hydraulic fracture when intersecting natural fracture gives rise to the development of a complex fracture network. A crossing model that is extended from the Renshaw-Pollard [12] interface crossing criterion, applicable to any intersection angle, has been developed, validated against the experimental data [16, 17], and was integrated at first in the UFM. The crossing model, showing good comparison with existing experimental data, did not account for the effect of fluid viscosity and flow rate on the crossing pattern. More recently a new advanced OpenT crossing model, taking into the account the impact of fluid and NF properties, have been developed [2] and integrated in UFM [18].

The modelling approach used in UFM to predict the leakoff into the NFs is presented below.

MODELING LEAKOFF INTO PERMEABLE NF IN UFM

The main assumptions for the current modelling of leakoff from HF into the intercepted NF in UFM are given below.

- The rock formation contains vertical discrete deformable fractures (NFs), which are initially closed but conductive because of their pre-existing apertures (due to surface roughness, etc).
- The propagation direction of the hydraulic fractures is not affected (unless intercepted) by closed and not invaded natural fractures. The intercepted NF could affect HFN propagation even from closed parts (when shear slippage takes place)
- The natural fractures are assumed to contain pore space and are permeable.
- The original fluid inside the natural fractures and in the reservoir (oil, gas, water) is compressible and Newtonian.
- Fracturing fluid can be incompressible or compressible and its rheology can be Newtonian or power law.
- The rock material is assumed to be permeable and elastic.
- When intercepted by the main hydraulic fracture, a natural fracture may remain closed, while still being able to accept fracturing fluid, or may be mechanically opened by fracturing fluid pressure depending on the magnitude of the fracturing fluid pressure, confining stresses applied on natural fracture, and frictional properties of natural fracture.
- The flow inside NFs is assumed to be 1D.
- The natural fractures can be opened by fluid pressure that exceeds the normal stress acting on them and/or experience Coulomb type frictional slip.
- The original natural fracture has width w_0 and are filled with reservoir fluid with pressure equal to pore pressure $p_0 = p_{res}$.
- Fluid flow invaded into the natural fracture develops along NFs. Invaded fracturing fluid into the NF may reach the end of NF, break the rock, and start to propagate into the rock accordingly to previously implemented propagation rules (only if the NF is opened). Fracture re-initiation from other points along the NF other than its ends (offsets) is not modeled at this time.

Hydraulic fractures propagating in the rock are modeled in accordance with existing approach in UFM model. The Schematic of the complex HF interaction with permeable NF is shown in Figure 2and Figure 4.

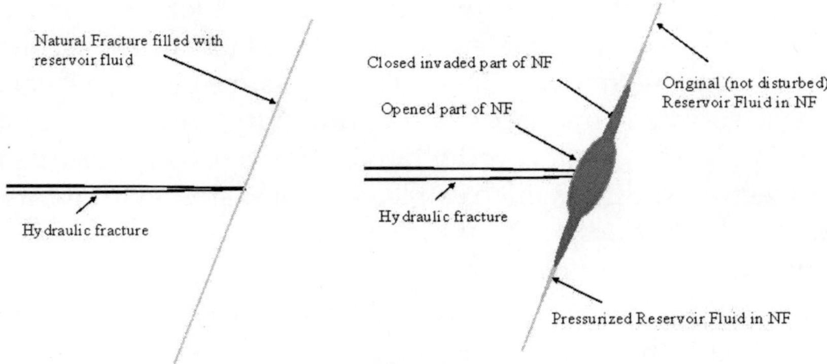

FIGURE 2.Hydraulic fracture intercepting natural fracture and possible situation to model

Fracturing fluid invasion into the two wings of the NF needs to be considered separately. Four possible regions can co-exist in each wing of the NF encountered by HF (Figure 4):

- Opened part filled with invaded fracturing fluid (fluid pressure exceeds the normal effective stress on NF), with length of opened part L opened >0
- Invaded closed part of NF (filtration zone) filled with fracturing fluid (fluid pressure above pore pressure but below the closure stress) with length L filtration >0
- Closed pressurized part filled with pressurized original reservoir fluid (fluid pressure above the pore pressure) with length L pressurized >0
- Closed undisturbed part of NF filled with reservoir fluid under original pore pressure conditions.

When a natural fracture is intercepted by the hydraulic fracture, the fluid pressure in the hydraulic fracture transmits into the natural fracture. If the fluid pressure is less than the normal effective stress on the natural fracture, the natural fracture remains closed. Even closed natural fractures may have hydraulic conductivities much larger than the surrounding rock matrix, and in this case fracturing fluid will invade the natural fractures more than leakoff into the surrounding matrix. If the portion of injected fluid is lost into closed natural fractures from the main HF, the HF growth could be affected.

For a closed fracture, the equivalent fluid conductivity is expected to change with the fluid pressure since contact deformation is a function of effective normal stress. This pressure-induced dilatancy and the associated increase in conductivity are important in increasing leak off. Also, any reduction in effective contact stress may result in fracture sliding, which can lead to local stress variations and slip induced fracture dilation, which can in turn change the overall conductivity of fracture networks.

The governing processes in first three regions listed above should be modeled, and the modeling approaches in different regions (also referred to as zones in the following context) are different due to different flow behaviors and rock/fluid properties.

BASIC GOVERNING EQUATIONS

Continuity Of Fluid Volume (Mass)

The equation for the continuity of incompressible fluid volume has the form

$$\frac{\partial q_{NF}}{\partial s} + \frac{\partial A}{\partial t} + q_L = 0, \qquad q_L = \frac{2hC_{tot}^{rock}}{\sqrt{t - \tau(s)}}, \quad A = \varpi h \tag{1}$$

where

q $NF^{(t)}$ - volumetric flow rate through a cross section of area A of natural fracture [m³/s]

A - cross sectional area of the natural fracture

q L - the volume rate of leakoff per unit length

□- average hydraulic fracture width (different from w, fracture opening by fluid pressure exceeding normal stress)

h - fracture height

C_{tot} - total leakoff coefficient from the wall of natural fracture

More generally in the case of compressible fluid the equation (1) should account for fluid density ρ f and mass flux q m. Considering the rate of change of fluid mass per unit length in a fracture

m˙, continuity of fluid mass in the fracture is governed by equation

$$\frac{\partial q_m}{\partial s} + \dot{m} + \rho_f q_L = 0 \qquad (2)$$

or along the fracture of constant length

$$\frac{\partial q_m}{\partial s} + \frac{\partial(\rho_f \varpi h)}{\partial t} + \rho_f q_L = 0, \qquad q_L = \frac{2hC_{tot}^{rock}}{\sqrt{t-\tau(s)}} \qquad (3)$$

Here s is coordinate along NF, and total leak off coefficient from the walls of the natural fracture C_{tot}^{rock} is equal to combined leak off coefficient [41]

$$C_{tot}^{rock} = C_{vc}^{rock} = \frac{2C_v^{rock} C_c^{rock}}{C_v^{rock} + \sqrt{C_v^{rock2} + 4C_c^{rock2}}} \qquad (4)$$

Where leak off coefficient for the filtration zone in the rock and leak off coefficient for the reservoir zone as shown in equations (5a - 5b)

$$C_v^{rock} = \sqrt{\frac{k_r \varphi_r \Delta p}{2\mu_f}} = \sqrt{\frac{k_r \varphi_r \Delta p}{2\mu_f}}, \ \Delta p = p_f - p_s \qquad \text{a}$$

$$C_c^{rock} = \sqrt{\frac{k_r \varphi_r c_T}{\pi \mu_r}} \Delta p, \quad \Delta p = p_f - p_r \qquad \text{b} \qquad (5)$$

$$\bar{C}_v = \frac{2(C_v)^2 \sqrt{t}}{V_L} \qquad \text{c}$$

with

ϕ_r- reservoir porosity

c_T- total compressibility of reservoir

k_r- permeability of rock matrix

μ_r- reservoir fluid viscosity in the porous media

μ_f- filtrate fluid viscosity

ρ_f- filtrate fluid density

p_r- reservoir pressure

In the case of multiple fluids, the invaded zone can be described by replacing

C_v with equivalent term (see (5c)) [42] Where

$\overline{C_v}$ is calculated using the average viscosity and relative permeability of all the filtrate fluids leaked off up to the current time, and V_L is the fluid volume per unit area that previously leaked off into the reservoir.

PRESSURE DROP ALONG CLOSED NF

The pressure drop along closed NF can be expressed from Darcy's law

$$q_{NF} = -\frac{k_{NF} A}{\mu_f} \frac{\Delta p}{L(t)}, \; A = \varpi h \tag{6}$$

Or for mass flux

$$q_m = -\rho_f \frac{k_{NF}}{\mu_f} A \frac{\partial p}{\partial s} \tag{7}$$

Then the pressure can be calculated from

$$\frac{\partial p}{\partial s} = -\frac{\mu_f}{\rho_f k_{NF} A} q_m = -\frac{\mu_f}{\rho_f k_{NF} \varpi h} q_m, \quad \text{at the inlet}: \; p = p_{in}(t) \tag{8}$$

Here

k_{NF} - permeability of natural fracture

μ_f - filtrate fluid viscosity

ρ_f - filtrate fluid density

A - cross sectional area of closed NF

p_{in} - fluid pressure at the inlet

CHANGE OF NF PERMEABILITY DUE TO STRESS AND PRESSURE CHANGES

The leak off into the natural fracture, or permeability of the natural fracture, is highly pressure dependent when pressure of invading fluid exceeds reservoir pressure but still is below the closure pressure. In general, permeability of natural fracture is a function of normal stress on NF, shear stress (or shear displacement due to shear slippage), and fluid pressure, and can be represented as a combination of permeability due to normal stress and permeability due to shear slippage [26]:

$$\begin{aligned} k_{NF} &= f(k_{NF}^n, k_{NF}^s) \\ k_{NF}^n &= f_1(k_o, \sigma_n, p\ ), \quad k_{NF}^s = f_2(u_s, \phi_{dil}) \\ k_{NF}^n &= k_o \left\{ C \ln\left[\frac{\sigma^*}{\sigma_n - p}\right]\right\}^3 \end{aligned} \tag{9}$$

Where constants C and σ* (reference stress state) are determined from field data, k o is the initial NF permeability (reservoir permeability under in-situ conditions), σ n is the normal stress on the NF, p is the pressure in the NF, and u s is shear-induced displacement (slippage).

THE WIDTH OF CLOSED INVADED NF

The width ϖ of closed NF invaded by the treatment fluid (hydraulic aperture) is related to the pressure-dependent permeability as [10]

$$\varpi = \sqrt{12k_{NF}} \tag{10}$$

The hydraulic width

ϖ

can be evaluated from Barton-Bandis model following approach [11,36,38], i.e. directly from Eq. (11) for given effective normal stress σ eff , reference effective stress σ_n^{ref} ,

, initial hydraulic fracture aperture ϖ_0 (related to the roughness of fracture surface), shear displacement us and dilation angle ϕ_{dil}

$$\varpi = \frac{\varpi_0}{1+9\frac{\sigma_{eff}}{\sigma_n^{ref}}} + \varpi_s + \varpi_{res}, \quad \varpi_s = |u_s| \tan(\phi_{dil}^{eff}), \quad \phi_{dil}^{eff} = \frac{\phi_{dil}}{1+9\frac{\sigma_{eff}}{\sigma_n^{ref}}} \tag{11}$$
$$\sigma_{eff} = \sigma_n - p_f(s)$$

SHEAR FAILURE

The second term in Eq. (11) represents the shear-induced dilation which contributes to the NF permeability (Eq.9). Shear induced dilation is related to the frictional slip which occurs when the shear stress reaches the frictional shear strength of the natural fractures

$$\tau_s = \lambda(\sigma_n - p)$$

. In the present study the NF propagation due to the shear induced slip is not considered, but the contribution of shear slippage zone to the NF enhanced permeability is evaluated based on the enhanced 2D DDM approach [40,43,44].

$$\sigma_n^i - p^i = \sum_j A^{ij} C_{nn}^{ij} D_n^j + \sum_j A^{ij} C_{ns}^{ij} D_s^j$$

$$\tau^i = \sum_j A^{ij} C_{sn}^{ij} D_n^j + \sum_j A^{ij} C_{ss}^{ij} D_s^j, \qquad A^{ij} = 1 - \frac{d_{ij}^{2.3}}{\left(d_{ij}^2 + h^2\right)^{2.3/2}} \tag{12}$$

The fracture surface slip (shear displacement) u_s can be found as shear displacement discontinuity D_s calculated for the closed sliding elements following Coulomb frictional law $\tau \geq \tau_s = \lambda(\sigma - p)$ from elasticity equations in the stress shadow calculation approach with accounting for the mechanical opening in HFN from Eq. (12).

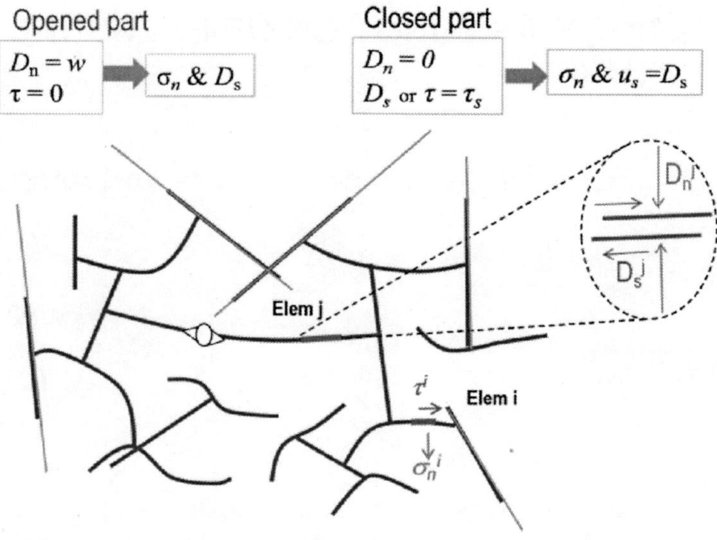

FIGURE 3. Stress Shadow Effect from opened (blue) HFN and closed parts (grey) of intercepted NFs

For any element j in the opened part of HFN (including opened part of intercepted NFs) the input normal displacement discontinuity D_n^j in Equation (12) is given by known fracture aperture (width) $D_n^j = w^j \neq 0$, and the shear stress is zero $\tau^j = 0$. Along the closed part of intercepted NFs the mechanical the opening is zero, $D_n^j = 0$, and the pressure and normal stress should be tracked to detect (find) elements sliding in shear. If $\tau^i \geq \tau_s^i$ then element i is sliding and dilating in shear, and the shear stress $\tau^i = \tau_s^i$ for this element is used to find fracture surface slip $u_s^i = D_s^i$ from the second equation (12). So, equation (12) could be solved for shear displacements D_s^j along opened and sliding in shear parts of total HFN and intercepted NF as schematically shown in Fig.3. In Eq.(12) C^{ij} are 2D, plane strain elastic influence coefficients [43] defining interactions between the elements i and j, and A^{ij} are 3D correction factors [44] accounting for the 3D effect due to fracture height h depending on the distance between elements d_{ij}.

FLUID DENSITY AS FUNCTION OF PRESSURE AND TEMPERATURE

Density of gas as a function of pressure and temperature has form

$$\rho_{gas} = \frac{p m_m}{ZRT} \tag{13}$$

Where p is pressure, T is temperature, Z is compressibility, R is gas constant, and m_m is molar mass. The changes in pressure and temperature with time produce changes in gas density. Density of a compressible fluid as function of pressure

$$\frac{d\rho}{dt}=\frac{\rho}{B}\frac{dp}{dt}=\rho c_f\frac{dp}{dt} \tag{14}$$

where B is bulk modulus (fluid elasticity) in Pa and $c_f=1/B$ is fluid compressibility in Pa^{-1}. More generally, change in fluid density due to changes in pressure and temperature for the low compressibility fluids through the known values of density ρ_0 and temperature T_0 at pressure p_0 has form (β here is volumetric expansion coefficient)

$$\rho=\frac{\rho_0}{1-\beta(T-T_0)}\times\frac{1}{1-\dfrac{p-p_0}{B}} \tag{15}$$

FLUID FLOW IN OPENED NF

The fluid flow in the opened NF ($p_f>\sigma_n$) will be handled as fluid flow in HFN and have been described before [1] depending on the flow regime:

Laminar fluid flow: Poiseuille Law [45]

$$\frac{\partial p}{\partial s}=-\alpha_0\frac{1}{\overline{w}^{2n'+1}}\frac{q}{H_{fl}}\left|\frac{q}{H_{fl}}\right|^{n'-1} \tag{16}$$

$$\alpha_0=\frac{2K'}{\varphi(n')^{n'}}\left(\frac{4n'+2}{n'}\right)^{n'};\quad \varphi(n')=\frac{1}{H_{fl}}\int_{H_{fl}}\left(\frac{w(z)}{\overline{w}}\right)^{\frac{2n'+1}{n'}}dz$$

Where

$\overline{w}$

is average fracture opening, and n' and K' are fluid power law exponent and consistency index.

Turbulent fluid flow ($N_{Re}>4000$):

$$\frac{\partial p}{\partial s}=-\frac{f\rho_f}{\overline{w}^3}\frac{q}{H_{fl}}\left|\frac{q}{H_{fl}}\right|, q=H_{fl}\left(-\frac{\overline{w}^3}{f\rho_f}\frac{dp}{ds}\right)^{1/2} \quad (17)$$

With Reynolds number (N_{Re}) for the power law fluid between parallel plates and *Fanning* friction factor (*f*) defined as

$$N_{Re}=\frac{3^{1-n'}2^{2-n'}\rho V^{2-n'}\overline{w}^{n'}}{K'\left(\frac{2n'+1}{3n'}\right)^{n'}}, f\approx\frac{1}{16\left[\log_{10}\left(\frac{\varepsilon}{7.4\overline{w}}\right)\right]^2}; \quad (18)$$

Where V is fluid velocity and ε is surface roughness height.

Darcy fluid flow through proppant pack of height h

$$\frac{\partial p}{\partial s}=-\frac{q\mu_{fl}}{kh\overline{w}} \quad (19)$$

will take place if the height of the fluid in fracture element become smaller than minimum fluid height. The minimum fluid height is calculated from the condition that pressure drop is equal to pressure drop due to the Darcy flow. The minimum height for turbulent and laminar flow is defined as

$$\text{Laminar flow:} \quad h_{fl}^{\min}=\left(\frac{k\alpha_0(n')H_{fl}}{\overline{w}^{2n'}\mu_{fl}}\right)^{1/n'}\frac{1}{q^{\frac{1}{n'}-1}}$$

$$\text{Turbulent flow:} \quad h_{fl}^{\min}=\left(\frac{kH_{fl}f\rho q}{\overline{w}^2\mu_{fl}}\right)^{1/2} \quad (20)$$

Where

μ_{fl}

is fluid dynamic viscosity, and k is proppant pack permeability. The boundary conditions at the inlet and tip of opened fracture

$$p = p_f(t), \quad p_{tip} = \sigma_n^{tip}(t) \tag{21}$$

where p_f is known fluid pressure at the intersection with natural fracture.

COMBINED FLUID FLOW INTO THE OPENED AND CLOSED PARTS OF INVADED NF

As I mentioned before, natural fracture can be closed, closed but invaded with fracturing fluid, closed and filled with pressurized reservoir fluid, or opened (Figure 4). The partially opened NF can contain opened, invaded, pressurized and closed parts which are dynamically changing with time. When fronts (positions of the boundaries between co-existing parts in invaded NF) change or propagate, the velocity of each propagating front can be considered as velocity of the corresponding fluid front in the relevant part of the invaded NF.

To properly model invasion of fracturing fluid into the NF, the propagation of each front should be modelled, and in different parts of the invaded NF different governing equations should be satisfied.

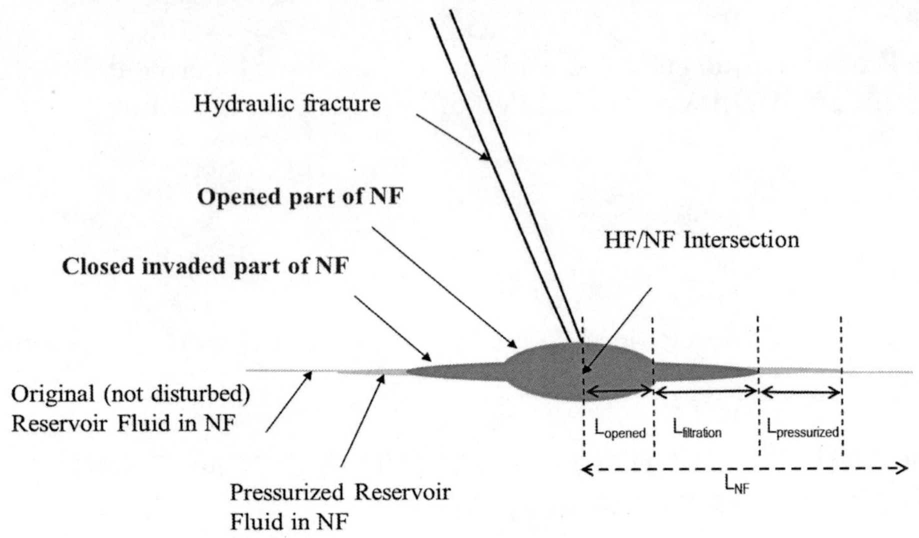

FIGURE 4. Details on different possible zones in intercepted permeable NF

OPENED PART OF NF

If the NF is opened at intersection element i with HF, then fluid pressure at intersection exceed the local normal stress on NF:

$$p_f(i) > \sigma_n^{NF}(i) \tag{22}$$

The tip of opened part of NF or the intersection element with HF (if NF is closed) becomes the inlet (injection point) into the closed invaded part of NF (filtration zone)

pfiltrin=popentip=σNFn

. Mention that if NF is completely opened, then it is a part of total hydraulic fracture network (HFN) and handled based on the approach described previously in [1], see also equations (16)-(21).

If NF is closed at intersection element i with HF, then fluid pressure is below the local normal stress, but can still be higher than reservoir (pore) pressure

$$p_o < p_f(i) \le \sigma_n^{NF}(i) \tag{23}$$

FILTRATION ZONE (CLOSED PART OF NF INVADED BY FRACTURING FLUID)

The fluid pressure, width and flow rate along the filtration zone can be calculated from given pressure $p_{in}^{filtr} = p_f(i)$ which satisfies condition (23). The flow rate along filtration zone can be iteratively solved from the system of equations (24) with flow rate from the inlet to filtration zone q_{in}^{filtr}

which is a part of total solution

$$\begin{cases} \dfrac{\partial p}{\partial s} = -\dfrac{\mu_f}{k_{NF}\varpi_{filtr}h}q, \quad \text{at the inlet: } p = p_{in}^{filtr}(t) \\ \dfrac{\partial q}{\partial s} + \dfrac{\partial(\varpi_{filtr}h)}{\partial t} + q_L = 0, \, q_L = \dfrac{2hC_{tot}^{rock}}{\sqrt{t-\tau_o(s)}} \\ k_{NF} = \dfrac{\varpi_{filtr}^{\;2}(p)}{12} \\ \varpi_{filtr}(p) = \dfrac{\varpi_0}{1+9\dfrac{\sigma_{eff}}{\sigma_n^{ref}}} + \varpi_s + \varpi_{res}, \quad \varpi_s = |u_s|\tan(\phi_{dil}^{eff}), \quad \sigma_{eff} = \sigma_n - p(s), \end{cases} \tag{24}$$

The length of filtration zone can be calculated by the tracking the volume of fracturing fluid leaked into the NF by marching from the inlet along the NF. At the end of filtration zone mass balance should be satisfied and fluid pressure will be higher or equal to the reservoir pressure

$$q_{in}^{filtr}dt = dVol_{frac}^{filtr}(dt) + dVol_{leak}^{filtr}(dt)$$
$$p_r < p^{end\ filtr\ zone} < \sigma_n^{NF} \qquad (25)$$

The flow rate (filtration/ pressurized front velocity) at the last element of filtration zone is used for calculations in the pressurized zone. The position of the front between the filtration zone and pressurized zone should be tracked, giving velocity of filtration front and pressure

p_{in}^{pres}

as input for solution in pressurized zone.

If NF is partially opened, then in the solution scheme for the filtration zone the intersection (inlet) element is replaced by the tip element i of the opened part of NF with

p(i)= σ_n^{NF} (i)

CLOSED PRESSURIZED PART OF NF (FILLED WITH RESERVOIR FLUID).

Mention that if the leak off coefficient for filtration zone $C_v^{rock} = 0, then\ p_r = p_f^{end\ filtr\ zone} < \sigma_n^{NF}$ and there will not be pressurized zone in

$$\text{NF} \left(p_r = p_f^{end\ filtr\ zone} < \sigma_n^{NF},\ L_{pressurized} = 0 \right).$$

For the general case of compressible reservoir fluid and non-zero total leak off from the walls of pressurizes NF into the rock, the leak off to the rock from the NF part filled with pressurized reservoir fluid is defined by the compressibility controlled leak off coefficient

$$C_c^{rock} = \sqrt{\frac{k_r \varphi_r c_T}{\pi \mu_r}} \Delta p, \quad \Delta p = p_{NF} - p_r \qquad (26)$$

The governing equations to calculate fluid pressure, width, and flow rate along the pressurized zone from known influx q filtr/ pres and pressure p filtr/pres are similar to Equations (24) for filtration zone with replacing fracturing fluid with reservoir fluid and using

ϖ_{pres}

as hydraulic width of pressurized zone of invaded NF, and c_T as reservoir fluid compressibility.

Notice that the pressure at the front between opened and filtration zones along invaded NF (or at HF/closed NF intersection point) and the time step are the inputs for the new pressure, width, and flow rate calculation in closed invaded part of intercepted NF. Initial influx q in can be prescribed based on the pressure in NF from the previous time step, and then the solution scheme will be applied from the intersection towards the end of NF with tracking the incremental mass balance (fluid injected to NF at current time step) to define the end of filtration front, and tracking pressure in the rest of NF (not invaded part, filled with reservoir fluid) to track the end of pressurized zone (fluid pressure equal to reservoir pressure). The flow rate and pressure are tracked and corresponding front positions are to be updated iteratively until in the pressurized zone of NF the following condition is satisfied (which indicates the position of the end of the pressurized zone)

$$\begin{aligned} q(L_{open} + L_{filtr} + L_{pres}) &= 0 \\ p(L_{open} + L_{filtr} + L_{pres}) &= p_r \end{aligned} \qquad (27)$$

At the end of pressurized zone pressure is equal to the reservoir pressure and flow rate is zero. If the end of NF is reached and pressure is above the reservoir pressure, then Equation (27) is replaced with condition q(L NF)=0. During pumping the pressurized zone extends until the end of NF, and then gradually shrinks, while the lengths of invaded zone and opened zones increase. Eventually whole NF will be filled with fracturing fluid, so the NF will contain only filtration and/or opened zones. When NF is completely opened, it becomes a part of total HFN. The elements in closed part of invaded NF can slip under some conditions (for example, due to the stress field change from stress shadow), influencing the calculations of total NF permeability k NF .

Each element in the closed part of intercepted NF is checked for shear slip possibility. The fracture surfaces slip (the shear displacement

u_s

) can be found as shear displacement discontinuity calculated for the closed sliding elements satisfying Coulomb frictional law

$\tau \geq \tau_{s=\lambda(\sigma-pf)}$

, from elasticity equations (12) accounting for the mechanical opening in HFN.

If some elements in closed not disturbed part of NF are sliding then the corresponding shear stress is involved in the stress shadow calculations and thus influences simulations results.

NUMERICAL APPROACH DESCRIPTION

The treatment of permeable natural fractures is a part of UFM model. It is possible to model leak off from HF into the NF form UFM in different ways, depending on the importance of required accuracy, numerical stability and CPU time.

The most computationally expensive but at the same time most accurate is a fully coupled numerical approach. The fully coupled approach means to discretise different parts of the invaded NFs and numerically solve the pressure as a part of total system of equations for the fracture network using iterative solution scheme. Another, more CPU efficient approach, is a decoupled numerical approach, where pressure and width along the invaded but closed part of NF are calculated separately based on the results from previous time step calculations along the HFN and corresponding pressure at HF/ NF intersections (inlets).

Two approaches have been considered to model interaction of hydraulic fracture with permeable natural fractures.

DECOUPLED NUMERICAL APPROACH

- When NF is intercepted by HF, create elements along whole NF to be used at next time step
- Evaluate initial guess of flow rate into the NF based on the pressure at the intersection and old pressure profile along closed NF
- Check for a possibility of frictional sliding along the closed parts of NFs to evaluate pressure-dependent permeability and conductivity along NF
- Iteratively calculate pressure, hydraulic width, flow rate and length of each zone along NF with using corresponding equations (for filtration and pressurized zones) by marching from intersection till the end of NF until condition (27) is satisfied indicating final results and final positions of zones' fronts
- Track the end of filtration zone for each pressure and flow rate iterations by checking the volume of fluid injected into NF at given time step
- Save invaded volume for volume balance, influx for mass balance, pressure at intersection, and time step to be used at the next time step
- Track the pressure at intersections and/or tips of opened HF part in NF to capture opened zones.
- If intersection is opened, use the pressure at the tip of opened HFN part along invaded NF for calculations along corresponding closed NF part
- Apply rules to treat special situations (intersecting NFs, etc)
- Save elements information (volume, pressure, flow rates) for the next time step

FULLY COUPLED NUMERICAL APPROACH

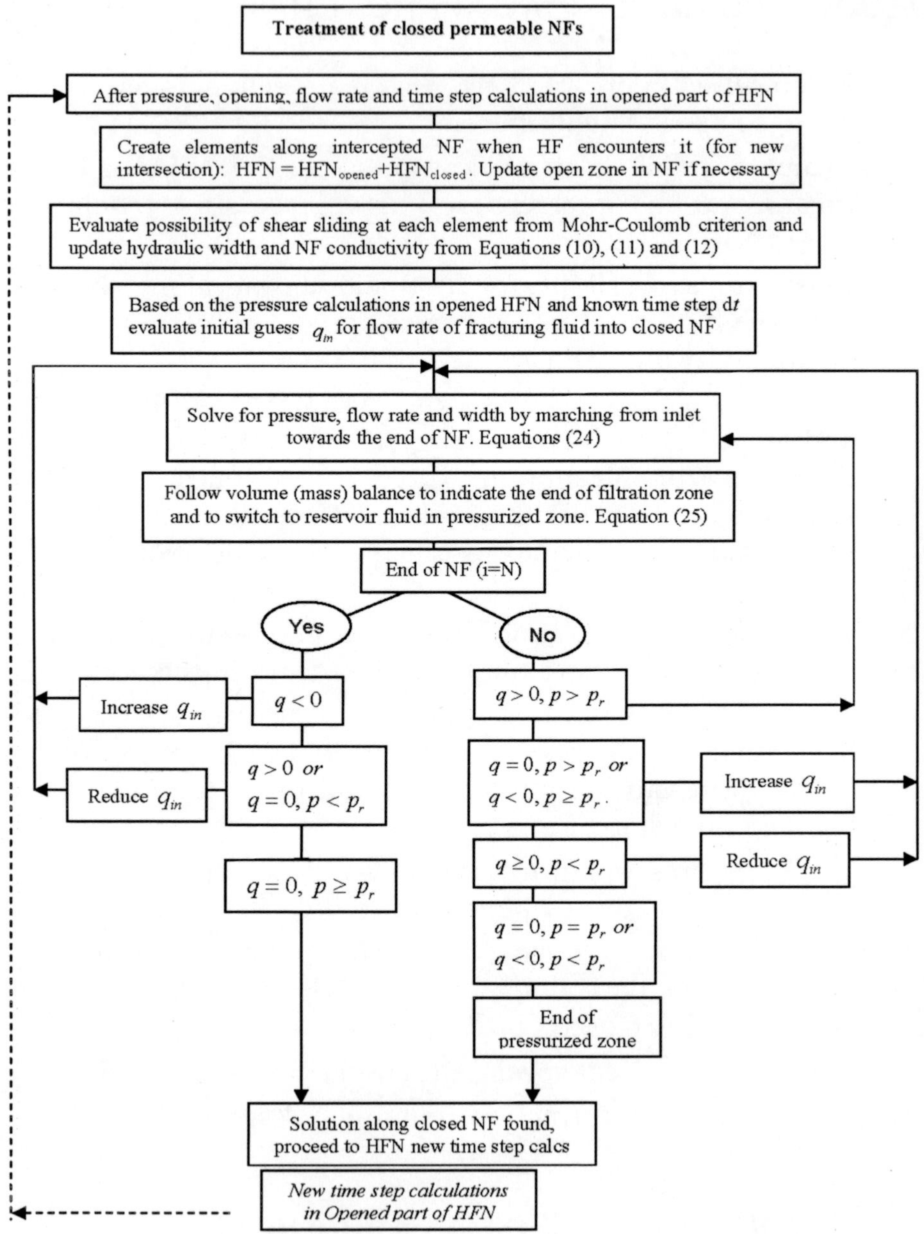

FIGURE 5. The proposed algorithm to account for leak off from HF into the permeable NFs (i=N indicates the end of NF)

Discretize the NF when HF intercepts it

- Make the NF elements a part of total network (HFN) and include pressure calculations along different zones of NF into the whole iterative scheme to calculate pressure, time step, and front positions.
- Calculate the flow rate and volume of fluid injected to NFs at each pressure iteration, and update/calculate pressure along HFN using existing scheme for opened elements and new equations (described above) for elements in closed invaded and pressurized NF parts
- From the calculated pressure in NF elements iteratively update the positions of the propagating fronts for opened, filtration, and pressurized zones in NF during pressure/time step iterations
- Track the pressure at intersections to capture when NF start to open
- Include the elements in the closed parts of invaded NFs into the stress shadow calculation scheme

The fully coupled NF modeling approach is heavier and more CPU expensive than decoupled approach. The Decoupled Numerical Approach has been selected as basic approach and it is described schematically on Figure 5 as a part of total solution

CONCLUSIONS

The new approach developed to account for the complex processes due to NF permeability accompanying HF/NF interaction have been presented in detail. This approach accounts for the important physical processes taking place during HF and permeable NF interaction, and will be implemented in UFM.

The next step is to evaluate the influence of leak off into the natural fractures during HFN simulations on the total HFN footprint and production forecast. The approach will be also validated against existing numerical, experimental and field data.

References

1. O Kresse, C. E Cohen, X Weng, R Wu, H Gu, Numerical modeling of hydraulic fracturing in naturally fractured formations. 45th US

Rock Mechanics/ Geomechanics Symposium, San Francisco, CA, 26 29June, 2011

2. D Chuprakov, O Melchaeva, R Prioul, Hydraulic Fracture Propagation Across a Weak Discontinuity Controlled by Fluid Injection. InTech; 2013
3. D. S Chuprakov, A. V Akulich, E Siebrits, M Thiercelin, Hydraulic Fracture Propagation in a Naturally Fractured reservoir. SPE 128715, Presented at the SPE Oil and Gas India Conference and Exhibition held in Mumbai, India, 20 22January, 2010
4. J Zhao, M Chen, Y Jin, G Zhang, Analysis of fracture propagation behavior and fracture geometry using tri-axial fracturing system in naturally fractured reservoirs. Int. J. Rock Mech. & Min. Sci, 45 2008 1143 1152
5. M Thiercelin, E Makkhyu, Stress field in the vicinity of a natural fault activated by the propagation of an induced hydraulic fracture. Proceedings of the 1st Canada-US Rock Mechanics Symposium 2007
6. X Zhang, R. G Jeffrey, The role of friction and secondary flaws on deflection and re-initiation of hydraulic fractures at orthogonal pre-existing fractures. Geophysical Journal International 2006 166 3 1454 1465
7. X Zhang, R. G Jeffrey, Reinitiation or termination of fluid-driven fractures at frictional bedding interfaces. J Geophys Res-Sol Ea. 2008Aug 28;113(B8).
8. X Zhang, R. G Jeffrey, M Thiercelin, Effects of Frictional Geological Discontinuities on hydraulic fracture propagation. SPE 106111, Presented at the SPE Hydraulic Fracturing Technology Conference, College Station, Texas, January29 31 2007
9. X Zhang, R Jeffrey, M Thiercelin, Deflection and propagation of fluid-driven fractures as frictional bedding interfaces: a numerical investigation. Journal of Structural Geology 2007
10. X Zhang, R. G Jeffrey, M Thiercelin, Mechanics of fluid-driven fracture growth in naturally fractured reservoirs with simple network geometries. Journal of Geophysical Research 2009B12406.
11. Beugelsdijk LJLde Pater CJ, Sato K. Experimental hydraulic fracture propagation in a multi-fractured medium. SPE 59419, presented at the SPE Asia Pacific Conference in Integrated Modeling for Asset Management, Yokohama, Japan, April 25 26 2000
12. C. E Renshaw, D. D Pollard, An Experimentally Verified Criterion for Propagation across Unbounded Frictional Interfaces in Brittle,

Linear Elastic-Materials. International Journal of Rock Mechanics and Mining Sciences & Geomechanics Abstracts. 1995Apr, 32 3 237 49

13. N. R Warpinski, L. W Teufel, Influence of Geologic Discontinuities on Hydraulic Fracture Propagation (includes associated papers 17011 and 17074). SPE Journal of Petroleum Technology. 1987 39 2 209 220
14. T. L Blanton, An Experimental Study of Interaction Between Hydraulically Induced and Pre-existing Fractures. SPE 10847, Presented at the SPE/DOE Unconventional Gas Recovery Symposium, Pittsburgh, PA, May 16 18 1982
15. T. L Blanton, Propagation of Hydraulically and Dynamically Induced Fractures in Naturally Fractured Reservoirs. SPE Unconventional Gas Technology Symposium; 01/01/1986; Louisville, Kentucky 1986
16. H Gu, X Weng, Criterion For Fractures Crossing Frictional Interfaces At Non-orthogonal Angles. 44th US Rock Mechanics Symposium and 5th US-Canada Rock Mechanics Symposium; 01/01/2010; Salt Lake City, Utah: American Rock Mechanics Association; 2010
17. H Gu, X Weng, J. B Lund, M Mack, U Ganguly, R Suarez-rivera, Hydraulic fracture crossing natural fracture at non-orthogonal angles, a criterion, its validation and applications. Paper SPE 139984 presented at the SPE Hydraulic Fracturing Conference and Exhibition, Woodlands, Texas, 24 26January, 2011
18. O Kresse, X Weng, D Chuprakov, R Prioul, C. E Cohen, Effect of Flow Rate and Viscosity on Complex Fracture Development in UFM. InTech; 2013
19. M. J Economides, K. G Nolte, Reservoir Simulation. Third edition. 2000
20. K. G Nolte, M. B Smith, Interpretation of Fracturing Pressures SPE 8297, September 1981
21. J. L Castillo, Modified Fracture Pressure decline Analysis Including Pressure-Dependent leakoff, SPE 16417, 1987
22. K. G Nolte, Fracturing Pressure Analysis for Non-Ideal Behavior. SPE 20704, JPT, February 1991
23. N. R Warpinski, Hydraulic Fracturing in Tight, Fissured Media. SPE 20154, JPT, February 1991
24. R. D Barree, H Mukherjee, Determination of Pressure Dependent Leakoff and its effect on Fracture geometry, SPE 36424, 1996. Presented at the 71st Annual Tech Conference and Exhibition, Denver Co, 6 9October, 1996
25. H Mukherjee, S Larkin, W Kordziel, Extension of Fracture Pressure Decline Curve Analysis to Fissured Formations. SPE 21872, 1991.

Presented at the Rocky Mountain Regional meeting and Low Permeability Reservoirs Symposium, Denver, Co, April 15 17 1991

26. J. B Walsh, Effect of Pore Pressure and Confining Pressure on Fracture Permeability, In. J. Rock Mech. Min. Sci. & Geomech. Abstr. 1981 18 429 435
27. N. R Warpinski, Fluid leakoff in natural fissures. In: Economides&Nolte: Reservoir Stimulation, 2000
28. B. R Meyer, L Bazan, A Discrete Fracture Network Model for Hydraulically Induced Fractures: Theory, Parametric and Case Studies, SPE 140514. Presented at the SPE Hydraulic Fracturing Tech Conference and Exhibition in The Woodlands, Texas, 2426 January 2011
29. S Rogers, D Elmo, R Dunphy, D Bearinger, Understanding Hydraulic fracture geometry and interactions in the Horn River Basin through DFN and Numerical modeling, SPE 137488, 2010. Presented at the Canadian Unconventional Resources & International Petroleum Conference, Calgary, Alberta, Canada, 19 21October, 2010
30. N Nagel, B Damjanac, X Garcia, M Sanchez-nagel, Discrete Element Hydraulic Fracture Modeling- Evaluating Changes in natural Fracture Aperture and Transmissivity, SPE 148957, 2011. Presented at the Canadian Resources Conference, Calgary, Alberta, Canada, Novebmer 15 17 2011
31. P Fu, S. M Johnson, C. R Carrigan, Simulating Complex Fracture Systems in Geothermal Reservoirs Using an Explicitly Coupled Hydro-Geomechanical model, ARMA 11 244Presented at 45th US Rock Mechanics/Geomechanics Symposium, Salt Lake City, UT, June 27-29, 2011
32. W. S Dershowitz, M. G Cottrell, D. H Lim, T. W Doe, A Discrete Fracture Network Approach for Evaluation of Hydraulic Fracture Stimulation of Naturally Fractured Reservoirs, ARMA-475, 2010. Presented at 44th US Rock Mechanics Symposium, San Francisco, CA, June 26 29 2010
33. M. M Rahman, A Aghigi, A. R Sheik, Numerical Modeling of Fully Coupled Hydraulic Fracture propagation in Naturally Fractured Poro-Elastic Reservoirs", SPE 121903, 2009. Presented at the 2009 SPE EUROPEC/EAGE Conference, Amsterdam, The Netherlands, 8 11June 2009
34. S. R Brown, R. L Bruhn, Fluid permeability of deformable fracture network. Journal of Geophys. Research 1998B2): 2489-2500.
35. M. L Cooke, C. A Underwood, Fracture termination and step-over at bedding interfaces due to frictional slip and interface opening. Journal

Structural Geology,2001 23 223 238

36. M. M Hossain, M. K Rahman, S. S Rahman, Volumetric Growth and Hydraulic Conductivity of naturally fractured reservoirs during hydraulic fracturing: A case study using Australian Conditions, SPE 63173, 2000. Presented at the 2000 SPE Technical Conference and Exhibition, Dallas, Texas, 1 4October 2000
37. D. A Chuprakov, A. V Akulich, E Siebrits, M Thiercelin, Hydraulic-Fracture Propagation in a Naturally Fractured Reservoir.SPE 128715, 2011. Presented at the SPE Oil and Gas India Conference and Exhibition, Mumbai, India, 20 22January 2010
38. K Tezuka, T Tamagawa, K Watanabe, Numerical Simulation of Hydraulic Shearing in Fractures Reservoir. Proceeding World Geothermal Congress, Antalya, Turkey, 24 25April 2005
39. X Weng, O Kresse, C Cohen, R Wu, H Gu, Modeling of Hydraulic Fracture Network Propagation in a Naturally Fractured Formation. Paper SPE 140253 presented at the SPE Hydraulic Fracturing Conference and Exhibition, Woodlands, Texas, USA, 24 26January, 2011
40. O Kresse, X Weng, R Wu, H Gu, Numerical modeling of Hydraulic fractures interaction in complex Naturally fractured formations. ARMA-292, Presented at 46th US Rock Mechanics /Geomechanics Symposium, Chicago, Il, USA, 24 27June 2012
41. R. H Dean, S. H Advani, An Exact Solution for Piston like Leak-off of Compressible Fluids. Journal of Energy Resources Technology 1983December, 106
42. A Settari, A New General Model of Fluid Loss in Hydraulic Fracturing, SPE 11625, August 1985 1985 491 501
43. S. L Crouch, A. M Starfield, Boundary Element Methods in Solid Mechanics. 1st ed. London: George Allen & Unwin Ltd, 1983
44. J. E Olson, Predicting fracture swarms: the influence of subcritical crack growth and the crack-tip process zone on joints spacing rock. In: Cosbrove JW, Engeder T (eds) The initiation, propagation and arrest of joints and other fractures. Geological Soc. Special Publications, London, 2004L 231 73 87
45. M. G Mack, N. R Warpinski, Mechanics of Hydraulic Fracturing. In: Reservoir Stimulation, Third Edition. Editors: Economides MJ and Nolte KG, 2000

Citations

CHAPTER 1

James Kear, Justine White, Andrew P. Bunger, Rob Jeffrey and Mir- Akbar Hessami (2013). Three Dimensional Forms of Closely-Spaced Hydraulic Fractures, Effective and Sustainable Hydraulic Fracturing, Dr. Rob Jeffrey (Ed.), ISBN: 978-953-51-1137-5, InTech, DOI: 10.5772/56261.

CHAPTER 2

F. Zhang, N. Nagel, B. Lee and M. Sanchez-Nagel (2013). Fracture Network Connectivity – A Key To Hydraulic Fracturing Effectiveness and Microseismicity Generation, Effective and Sustainable Hydraulic Fracturing, Dr. Rob Jeffrey (Ed.), ISBN: 978-953-51-1137-5, InTech, DOI: 10.5772/56302

CHAPTER 3

Brice Lecampion, Anthony Peirce, Emmanuel Detournay, Xi Zhang, Zuorong Chen, Andrew Bunger, Christine Detournay, John Napier, Safdar Abbas, Dmitry Garagash and Peter Cundall (2013). The Impact of the Near-Tip Logic on the Accuracy and Convergence Rate of Hydraulic Fracture Simulators Compared to Reference Solutions, Effective and Sustainable Hydraulic Fracturing, Dr. Rob Jeffrey (Ed.), ISBN: 978-953-51-1137-5, InTech, DOI: 10.5772/56212.

CHAPTER 4

Trinath Sahoo (2012). Strategies to Increase Energy Efficiency of Centrifugal Pumps, Centrifugal Pumps, Dr. Dimitris Papantonis (Ed.), ISBN: 978-953-51-0051-5, InTech, DOI: 10.5772/25870.

CHAPTER 5

Charles Fairhurst (2013). Fractures and Fracturing - Hydraulic fracturing in Jointed Rock, Effective and Sustainable Hydraulic Fracturing, Dr. Rob Jeffrey (Ed.), ISBN: 978-953-51-1137-5, InTech, DOI: 10.5772/56366.

CHAPTER 6

Norihisa Miki (2013). Liquid Encapsulation Technology for Microelectromechanical Systems, Advances in Micro/Nano Electromechanical Systems and Fabrication Technologies, Assistant Professor Kenichi Takahata (Ed.), ISBN: 978-953-51-1085-9, InTech, DOI: 10.5772/55514.

CHAPTER 7

Fabrício Gonzalez Nogueira, José Adolfo da Silva Sena, Anderson Roberto Barbosa de Moraes, Maria da Conceição Pereira Fonseca, Walter Barra Junior, Carlos Tavares da Costa Junior, José Augusto Lima Barreiros, Benedito das Graças Duarte Rodrigues and Pedro Wenilton Barbosa Duarte (2013). Design and Field Tests of a Digital Control System to Damping Electromechanical Oscillations Between Large Diesel Generators, Diesel Engine - Combustion, Emissions and Condition Monitoring, Dr. Saiful Bari (Ed.), ISBN: 978-953-51-1120-7, InTech, DOI: 10.5772/55707.

CHAPTER 8

Sau-Wai Wong, Mikhail Geilikman and Guanshui Xu (2013). The Geomechanical Interaction of Multiple Hydraulic Fractures in Horizontal Wells, Effective and Sustainable Hydraulic Fracturing, Dr. Rob Jeffrey (Ed.), ISBN: 978-953-51-1137-5, InTech, DOI: 10.5772/56385.

CHAPTER 9

Michael Molenda, Ferdinand Stöckhert, Sebastian Brenne and Michael Alber (2013). Comparison of Hydraulic and Conventional Tensile Strength Tests, Effective and Sustainable Hydraulic Fracturing, Dr. Rob Jeffrey (Ed.), ISBN: 978-953-51-1137-5, InTech, DOI: 10.5772/56300.

CHAPTER 10

Olga Kresse, Xiaowei Weng, Dimitry Chuprakov, Romain Prioul and Charles Cohen (2013). Effect of Flow Rate and Viscosity on Complex Fracture Development in UFM Model, Effective and Sustainable Hydraulic Fracturing, Dr. Rob Jeffrey (Ed.), ISBN: 978-953-51-1137-5, InTech, DOI: 10.5772/56406.

CHAPTER 10

Olga Kresse, Xiaowei Weng, Dimitry Chuprakov, Romain Prioul and Charles Cohen (2013). Effect of Flow Rate and Viscosity on Complex Fracture Development in UFM Model, Effective and Sustainable Hydraulic Fracturing, Dr. Rob Jeffrey (Ed.), ISBN: 978-953-51-1137-5, InTech, DOI: 10.5772/56406

CHAPTER 11

Olga Kresse and Xiaowei Weng (2013). Hydraulic Fracturing in Formations with Permeable Natural Fractures, Effective and Sustainable Hydraulic Fracturing, Dr. Rob Jeffrey (Ed.), ISBN: 978-953-51-1137-5, InTech, DOI: 10.5772/56446.

INDEX